MONEY, OWNERSHIP, AND AGENCY

A survey of the elementary functions of money and its invariant representations, alongside the complementary topic of ownership. The book uses the language of Promise Theory to explain the semantics and lifecycle of money in a concise and semi-formal manner, with the goal of reaching a modern probabilistic view of money and economics as a network, independent of philosophical or political treatments as well as models that build on Game Theory. Relativistic notions like 'value' are distinguished carefully from invariant measures of amount to decouple frequently muddled concepts in the popular literature, and money is distinguished from its many inequivalent representations (i.e. proxy technologies). A key strength of this work lies in the way it exposes money as an interconnection network, which has evolved to communicate intentions and decisions between agents in a network, where prices form a language for information channels that we call markets.

This book is aimed at economists, philosophers, and engineers. It introduces readers to the concepts in a practical manner, building on the concept of voluntary cooperation. The book draws on many examples from the real world, with a particular emphasis on semi-formal semantics.

A must read for anyone involved in designing intelligent digital currency.
–Mike Warner (Senior Strategist, San Francisco Federal Reserve)

'*Promise Theory offers a methodology for generating certainty on top of uncertain foundations. This book presents the formal foundations of Promise Theory. It lays out the formalisms in a clear, concise, understandable way that makes them accessible to non-mathematicians.*'
– Jeff Sussna, Author of Designing Delivery

Also by the authors:

PROMISE THEORY: PRINCIPLES AND APPLICATIONS

PROMISE THEORY: CASE STUDY ON THE 2016 BREXIT VOTE

Other reviews:

'A landmark book in the development of our craft...'
–Adrian Cockcroft (about In Search of Certainty)

'Proud to say that I am a card-carrying member of the [Mark Burgess] fan club. And I think that it wouldn't be too much of a stretch to say that he's the closest thing to Richard Feynman within our industry (and not just because of his IQ).' –Cameron Haight (about Smart Spacetime)

'...our whole industry is transforming based on ideas [Mark Burgess] pioneered'
–Michael Nygard (about Smart Spacetime)

'The work done by [Mark] on complexity of systems is a cornerstone in design of large scale distributed systems...'
–Jan Wiersma (about In Search of Certainty)

'Some authors tread well worn paths in comfortable realms. Mark not only blazes new trails, but does so in undiscovered countries.'
–Dan Klein (about Smart Spacetime)

MONEY, OWNERSHIP, AND AGENCY

AS AN APPLICATION OF PROMISE THEORY

JAN A. BERGSTRA AND MARK BURGESS

χtAxis press

CONTENTS

PREFACE

This text was originally written early in 2017, with the aim of using Promise Theory to partially formalize some of the functional semantics of money, in its many different forms. The material has been lightly edited and reformatted in book form, with an index, in order to make it more widely available and usable. It's a concise attempt to survey the elementary functions of money, from both a semantic and a dynamical perspective; it does not concern monetary policy or economic modelling, except for some brief remarks for context.

In these chapters, we probe the invariant representations for money alongside the inevitable and complementary topic of ownership. The language of Promise Theory helps to bring a concise order to the discussion (see [BB14a]), and offers a possible path to shift away from the differential style of economic modelling—whose assumptions and therefore applicability can only be satisfied on the timescale of decades—towards a more modern and probabilistic view, with a more human, day to day scale.

We've done our best to separate subjective notions like 'value' from invariant measures of money, and offer a simple formalization of common concepts. Although idealized, we believe this allows simple insights to emerge. We further distinguish the concept of money from its many inequivalent proxies, and show that the traditional relativistic notions of 'utility' and 'value' are not needed to understand monetary systems.

Money's principal strength seems to lie in its role as an *invariant representation* of an interconnection network, carrying agents' autonomous intentions and decisions. We show how prices act as an information channel to form markets—in the past a price has been little more than a number with simple semantics, but the price (marketing) channel is the easiest to extend using technology, and is likely to be enhanced significantly in the information age. A single currency or monetary system functions as a network transport agent, which can be routed freely through hubs (banks) or peer to peer (with cash), and which allows a complex balance of payments between more than two parties over time.

Finally, we show why it's advantageous to treat money as a conserved quantity, for accounting purposes, because this coincides with the goal of trust and fairness, but we also warn that money has no formal basis for conservation. The creation and destruction of money by banks does not compromise the integrity of money supply, provided certain semantics are upheld, but the practice of compound interest may have unpredictable results. The semantics of money may linger on after money technologies have disappeared, both in the form of a trust network, and in residual memory of contractual terms and conditions. In summary, we find that protocols for interaction matter at least as much as quantities and exchange rates, and we believe that Promise Theory is uniquely useful in contemplating such matters, before premature implementation takes over.

We've tried to be concise in formulation, without undue handwaving. Consequently, we only manage to scratch the surface of the subject. Nonetheless, we've found that Promise Theory is quite revealing of the simple mechanisms at work in economics and in day to day transactions. We hope readers agree.

Jan A. Bergstra
Mark Burgess
2019

CHAPTER 1

INTRODUCTION

We ask the straightforward question: how can money be described in a semi-formal but pragmatic manner, suitable for the modern age, and without getting bogged down in politics or philosophy? During the 20th century science rejected determinism as a fundamental paradigm, and embraced probabilistic methods. This transition seems to have escaped economics. If we are to make that transition, a reformulation of the subject, equipped with tools of sufficient formal sophistication is needed. Our goal is to begin with rather simple matters about the functional semantics of money, and the environment in which it operates, to map out the nature of its interactions.

The amount of work needed to give substance to even simple claims is quite considerable, as testified by the length of this book, but seems nonetheless doable. We have tried to make the level of detail plausible, while recognizing that the burden for readers to assimilate all the material is formidable. We thus ask the forbearance of readers who might be impatient to get to the 'bigger issues', and hope that bringing all this material together will be valuable. All rational description has to begin by scaling from the bottom up.

What, then, is money? The nature of money has changed slightly over recorded history. Is it information, a phantom of value or utility, or something else entirely? Standard lore suggests that money, goods, capital, and the other denizens of economic theory have more than just one role to play. We shall try to discover some of these without preconceived ideas, focusing mainly on the present, but taking care to incorporate remnants of the past[Gra11, Smi04]. One role in particular, for money, which turns out to be quite important to the future, is that of a technology for *networking* people, companies, goods, services, etc. This theme recurs several times, and has important implications. We attempt to develop this modern informational viewpoint.

Assumptions of money's existence and role are ubiquitous and inescapable (though when to call something money or not seems contentious). The definitions of buying, selling, and trading, in textbooks today are so impregnated with the concept of money that it is hard to find any reference that does not assume it as an interloper in transactions. Yet, as money changes in character, we need a way to understand it independently of these superficial appearances. The history of money is long and involved[Har00, Fer08, Har11, Bei05, Smi04, Suz17, Gra11]. It is a study in technologies for exchange. From the 19th century, economics was developed along side (and often in the image of) the physics [Mir89]. It is surely no accident that Newton worked at the Royal mint. Economics borrows freely of the concepts and language of continuous causal transfer, like the Newtonian mechanics. which predates the modern non-deterministic understanding of systems. The language of differential calculus is thus ubiquitous, yet this already conceals significant assumptions about scale, continuity, and the smoothness of change. The differentiable picture, with infinite accuracy, contrasts with the day to say transactional nature of the economy, where certain cashflow payments may be blocked for want of sufficient funds: money is everywhere discrete at the scale of human concerns, and even where numbers are rounded to the smallest convenient monetary unit. We also note, in passing, that non-linearities in any economic network may potentially lead to the uncontrolled amplification of these approximations.

One of the conspicuous missing pieces in economics seems to be the role of time. Money's usage cannot be discussed without reference to time, because its semantics are intrinsically linked to expiring intervals, duels, races, and shifting trends. This is part of a larger omission, which is proper theory of scales. The scale over which calculus might make sense, and the statistical smoothness bulk transactions (close to equilibrium) would seem to require averaging over several orders of magnitude larger than these transactional aspects of human intent. Could it be that present day economics does not address the scale of human concerns, but only models a hypothetical long-term limit that doesn't exist? These are the kinds of questions we eventually hope to be able to reframe more precisely from a closer attention to detail.

Our goal here, then, is to re-examine money as a network of intentional interactions, and to define the agents and interactions from information theoretic and semantic perspectives. On top of such considerations, we build the lowest level foundation of a formal economic theory, with a minimum of assumption, and aiming for the same degree of humility as that expressed by Von Neumann and Morgenstern about their application of Game Theory as a new paradigm[NM44]. Our main concern will be to determine what invariant properties can be divined in a logically consistent way, and may survive into the future as new technologies overtake our financial systems. Four aspects of money will be of particular interest:

- The mechanics of money.

- The subjective assessment of value.

- The role of price as a semantic network.

- The role of spacetime scales, and the nonlinear effects of 'interest'.

1.1 LITERATURE

The concepts of money and value have been mixed and muddled since the time of Aristotle[Sch54]. Modern economics, beginning with Adam Smith and Karl Marx, seemed entranced by the goal of discovering a universal notion of 'value' to justify the existence of money, egged on by the successes of applying universal invariances in physics[Mir89]. Utility theories have since compounded this issue, suggesting a level of rational behaviour that has largely been debunked. Indeed, from the present study, it seems to be money's detachment from subjective 'value' that affords it the status of an *invariant measure*, and this (without contradiction) is what allows it to measure value, according to anyone's favourite standard, through the notion of price. For the concept of value to retain meaning, as it exists in the minds of individuals, it is essential to retain its status as a variable assessment, and rather to look for invariant aspects elsewhere—just as one acknowledges a distinction between the concept of length and that of a metre. We cover these issues in later chapters.

For industrial age Marx, 'value' originated pragmatically in the abstraction of human labour, and thus began a long standing attachment between value, wages, prices, and markets. The philosopher Simmel wrote his treatise The *Philosophy of Money* in 1900 [Sim11] speaks of an age before the complexities of modern financialization, with a pedestrian viewpoint that makes no distinction between value, utility, and price. Simmel simply equates these, claiming: 'Value is, so to speak, an epigone of price and the statement that they must be identical is a tautology'. The reason, he claims, is that money, as an interchange language, becomes a plausible surrogate for an objective notion of value. This final assertion seems quite plausible to us, as it relieves consumers of the burden of assessing value for themselves, choosing rather to trust in third parties to take the lead. However, the equation of price and value is manifestly wrong according to our work. Simmel identified the depersonalization of relationships through an intellectualization of money as a shift towards materialism. It is possible that this cultural shift could yet be undone in the age information technology.

A compelling attempt to confront a definition of 20th century money can be found in Keynes *Treatise on Money*. This treatise was quickly overshadowed by *General Theory of Employment, Interest and Money* where Keynes applied it to the matter of the Great Depression[Key30, Key64]. The latter drew economists' attentions away from basic matters of definition, back to how prices and money might be determined by monetary policy and interest rates, laying the foundations for competing theories of macroeconomic equilibria[NM44, Mir89, Kee11, BG02, Sal05], which were later disrupted by the highly politicized monetary theory of Friedman[Fri02] and Hayek[Hay44]. The role of money as a neutral form of memory and trust in society has also been explored by economic anthropologists and social scientists[Har00, Har07, Har05].

1.2 SCOPE

We begin without attachment to these prior discussions, building from the basis of a network of cooperating agents. We formalize our assumptions on Promise Theory[1], which describes agents that interact non-deterministically, but which advertise their intentions and properties as information. The relationship between promises and trust is also well understood[BB06b]. Promise Theory is a theory of voluntary cooperation [BB14a], with a well defined theory of scaling [Bur14, Bur15, Bur16b]). How can Promise Theory help? In fact, it can help in reformulating the basic theory of money at a level of necessary and sufficient formality to be able to ask precise questions. In particular, we are interested in asking questions like the following:

- What are the agents involved in money? What are their promises?

- Can we define how to count monetary quantities in terms of agents and their promises.

- What is the relationship between money and property?

- What dependencies are implicit in the use of money?

- Does the source of money (i.e. how it was created) matter?

- Can a monetary system be sustained over time?

- What is the relationship between money, location, and time?

- Is there a limit on the number of currencies that can exist?

- Does the technology used to represent money (i.e. act as its proxy) matter?

To keep our description self-contained, we restrict ourselves to the basic principles of Promise Theory, as described in [BB13, BB14b, BB14a][2]. Readers unfamiliar with Promise Theory can imagine a description based on sets for describing measures and networks for describing interactions. These are briefly summarized below.

1.3 REMARKS ABOUT THE USE OF PROMISE THEORY

Our use of Promise Theory might be unsettling for some readers. In many ways, like money, Promise Theory has the status of a specialized *lingua franca* for discussion relationships. In Promise Theory, everything is an *agent*, a *promise*, an *imposition*, or an *assessment*. Implicitly, there are also *events*, though these are taken more for granted in promise literature. This kind of abstraction will not seem unusual to physicists or mathematically inclined readers, but to those from economics or the social sciences it could be challenging to accept such rigid terminology with specific meanings. By using the concept of a promise so universally, we risk devaluing it in the eyes of some (as physicists devalue the concept of 'particle', for instance).

It is a common complaint about Promise Theory that inanimate, and even virtual things (like data), would be considered able to promise anything. Is a promise not something only humans can offer? However, there is no contradiction. This is easily countered by the following: it is very common language to say 'the table promises to be sturdy' or 'the weather promises to be fine' (see the discussion in [BB14a]). These appear as intentional and semantic statements that harmlessly project human interpretations onto inhuman phenomena[3]. Thus we understand that it is in the nature of semantics (an intrinsically human phenomenon) that these objects or phenomena make effective promises by proxy. Intentions are human, but non-human things can also make promises because of the network effect, agents acting as 'modular concerns'. We are not directly addressing issues of psychology, morals or ethics. The extent to which these matters appear is only through the appearance of certain promises.

If one can overcome the unfamiliarity of the abstraction, then (like money) Promise Theory opens a door to discuss almost anything in a rational framework, with a small number of sound principles to guide the discussion. All of this, we claim, is more than ample reason to use the promise abstraction.

1.4 ASPECTS OF PROMISE THEORY WE NEED

Promise Theory begins with the idea of autonomous agents that interact through the promises and impositions they make to one another. Promises are declarations of intent about the agent making the promise, while impositions are an attempt to induce an intent in another agent (to impose upon them, without prior warning)[Bur05a, BB14a]. Agents have no a priori structure. We allow them to exhibit the appearance of agency or intent, either fundamentally or by proxy. This means that an observer would interpret their behaviours as being intended and these semantics are ultimately always assessed by the observer. Agents are autonomous in the sense that they govern their own behaviour, Each agent's promises are made by itself (or channelled as proxy on behalf of another), and other agents' attempts to make promises on its behalf may be assumed ineffective, unless it promises to subordinate itself to external command. Each agent is thus responsible for keeping the promises it makes, and not reasonably responsible for keeping promises others might make on its behalf.

Our view on promises is distinct from others, in that: i) promises do no to create legally enforceable tasks or objectives, ii) promises do not create moral or ethical bindings. These caveats taken together is phrased as: promises do not create obligations[4]. What promises do, however, is: iii) modify expectations (by promisee and other agents in scope) about the promised state of affairs, iv) to prepare for modified assessments of trust and other valuations in the promiser. A brief example, of how promises relate to (but are independent of) obligations, in our viewpoint, helps to clarify this.

Example 1. *Suppose A_1 promises to pay to A_2 amount μ upon receiving good G from A_2, and A_1 promises to accept a legally enforceable penalty P upon not paying amount μ to A_2 within 10 days after receiving G from A_2. These two promises together construct an enforceable obligation (contract) on the side of A_1. The promising of a legally enforceable obligation to perform (deliver) G to A_2 is a legally binding self-imposition. In the same way A_1 can promise to have a morally binding (but not legally enforceable) obligation to provide G for A_2. This is a morally binding self-imposition. Together these conventions create a setting in which this version of Promise Theory provides the most versatile concept of a promise, with limited impact on alternative philosophies at large.*

Our notation is as follows: we write an autonomous promise from Promiser to Promisee, with body b as:

$$\text{Promiser} \xrightarrow{b} \text{Promisee}, \tag{1.1}$$

and we denote an imposition by

$$\text{Imposer} \xrightarrow{b} \blacksquare \text{Imposee}. \tag{1.2}$$

Promises come in two polarities, denoted with a $\pm$ signs, as below. The $+$ sign gives assertion (offer) semantics:

$$x_1 \xrightarrow{+b} x_2 \quad (\text{I offer } b) \tag{1.3}$$

while the $-$ sign gives projection (acceptance) semantics:

$$x_1 \xrightarrow{-b} x_2 \quad (\text{I will accept } b) \tag{1.4}$$

where x_i denote autonomous agents. A promise to give or provide a behavior b is denoted by a body $+b$; a promise to accept something is denoted $-b$ (or sometimes $U(b)$, meaning use-b). Similarly, an imposition on an agent to give something would have body $+b$, while an imposition to accept something has a body $-b$. In general, intent is not transmitted from one agent to another unless it is both $+$ promised and accepted with a $-$. Such neutral bindings are the exchange symmetry.

A promise model thus consists of a graph of vertices (*agents*), and edges (either *promises* or *impositions*) used to communicate intentions. Agents publish their intentions and agents in scope of those promises may or may not choose to pay attention. In that sense, it forms a chemistry of intent [Bur13a], with no particular manifesto, other than to decompose systems into the set of necessary and sufficient promises to model intended behavior.

A promise binding defines a voluntary constraint on agents. The perceived strength of that binding is a value judgement made by each individual agent in scope of the promises. If an agent offers b_1 and another agent accepts b_2, the possible overlap $b_1 \cap b_2$ is called the effective action of the promise.

For example, A promises B 'to give an apple'. This does not imply that B will accept the apple. B might then promise A to 'accept an apple'. Now both are in a position to conclude that there is a non-zero probability that an apple will be transferred from A to B at some time in the future, nothing more. If the promise is to continuously transfer apples, then the timing is less ambiguous.

The constraints implied by the scope of observability for agents complicates this. Consider an exchange of promised behaviour, in which one agent offers an amount b_1 of something, and the recipient promises in return to accept an amount b_2 of the promised offer.

$$\pi_1 : x_1 \quad \xrightarrow[\sigma_1]{+b_1} \quad x_2 \qquad (1.5)$$

$$\pi_2 : x_2 \quad \xrightarrow[\sigma_2]{-b_2} \quad x_1 \qquad (1.6)$$

Then any agent in scope σ_1 of promise π_1, will perceive that the level of promised cooperation between x_1 and x_2 is likely b_1. An agent in scope σ_2 of promise π_2, will perceive that the level of promised cooperation between x_1 and x_2 is likely b_2. Finally, an agent in scope $\sigma_1 \cap \sigma_2$ of both promises π_1 and π_2, will perceive that the level of promised cooperation between x_1 and x_2 is likely $b_1 \cap b_2$. Promises are assessed by each and every agents individually. The relativity of observations can lead to peculiar behaviours, contrary to expectation. Ultimately every agent makes decisions based on the information it has.

If a promise with body S is provided subject to the provision of a pre-requisite promise π, then the provision of the pre-requisite by an assistant is acceptable if and only if the principal promiser also promises to acquire the service π from an assistant (promise labelled $-X$):

$$x_T \xrightarrow{+b(\pi)} x_1, \quad \left.\begin{array}{c} x_1 \xrightarrow{S|b(\pi)} x_2 \\ x_1 \xrightarrow{-b(\pi)} x_2 \end{array}\right\} \sim x_T \xrightarrow{+b(\pi)} x_1, x_1 \xrightarrow{S} x_2 \qquad (1.7)$$

The relativity of observers and their assessments is the key to understanding a local agent view of behaviour. Intent, being an interpretation offered by an observer, brings with it a variety of anthropomorphisms, like trust and level of belief which are equally important to science (witness the Bayesian interpretation of statistical observations for instance). This should not be considered a problem; it is only the reflection of a received interpretation by local observers. Promise Theory, like statistics and quantum mechanics, is a theory of incomplete information. The promise formalism is described in[BB14a].

Definition 1 (Axioms for reasoning). *The following labels are used in our promise clusters to show necessary conditions:*

1. (**Ax1**) *All promises and impositions are made autonomously by their 'promiser' agent.*

2. (**Ax2**) *A complementary $\pm b_i$ pair of promises, between agents has the probable effect $b_+ \cap b_-$. If either polarity in the pair is missing, the promise is ineffective.*

3. (**Ax3**) *A conditional promise is only effective if the promiser also promises the condition itself, or promises to accept the condition from another agent, referring to the construction in (1.7).*

4. (**Ax4**) *All assessments are local to the agent making them (subjective), i.e. assessments cannot be agreed upon without mutual promises to reach an equilibrium.*

By 'effective' we refer to the likelihood of them bringing about the promised outcome.

Finally, the central aspects of Promise Theory we shall rely on may be found summarized in the sections of [BB14a]:

1. The axioms and definitions of promises and impositions, chapter 3.

2. Promise assessment, section 5.2 for keeping track of causal influence

3. Conditional and assisted promises, section 6.2.

4. Agreements, section 8.4

5. The value of a promise, chapter 9.

6. Specification of timescales for the lifecycle of a promise, section 7.1.

We shall also refer to some of the scaling results from [Bur14, Bur15]. Readers may find it helpful to refer to the breakdown of concepts into the three poles of Promise Theory, shown in figure 1.1.

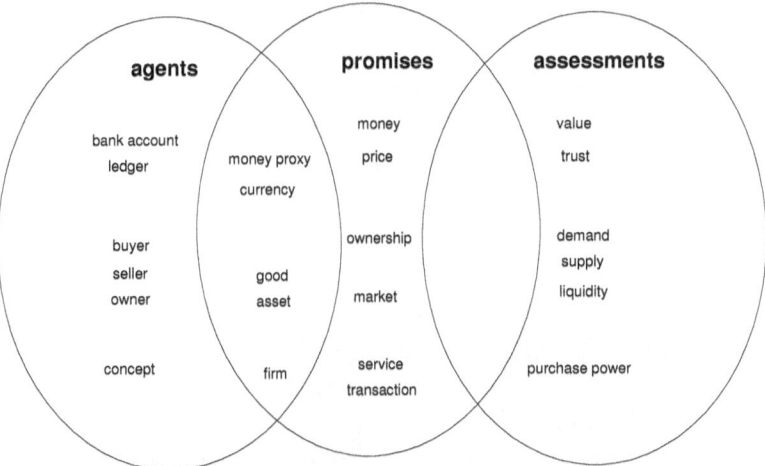

Figure 1.1: A classification of monetary subjects into Promise Theory subjects: agents (which are part of promises), promises (specifically the body of subject of the promise) and assessments about the status of the promise by different parties. Things.

CHAPTER 2

TRUST AND VALUE

Promises are very useful as a parsimonious representation of agents interacting. They allow us to replace many different assessments, based on different criteria with a simple notion of trust[BB06b]. Trust acts as a universal currency of its own. The penalty for not keeping a promise lies in the accounting of trust between the parties. Withdrawal of a promise (unpromising) is an inverse of promising. Withdrawal of a promise impact expectations and trust. Withdrawal may decrease trust if it is perceived as an announcement that a promise won't be kept. However, if the promisee has already worked out that the probability of the promise being kept has become very low it might even increase trust! The promiser, after all, owns up to this by withdrawing the promise. If coins or banknotes are not anymore valid, they must withdraw the promises they make.

Unlike the notion of utility in some social theories, Promise Theory trust evolves and emerges over time, as in Axelrod's iterative games[Axe97, Axe84]. It has no objective or pre-ordained deterministic value, as is often attributed to value or wealth. It evolves as a cognitive learning process[Bur16b, Bur17b], and leads to a set of partial orderings of agents, in the eyes of each agent, as a kind of weight or preference for interacting. It thus cements the connection between assessment and memory. Agents who cannot remember anything also cannot modify their assessment of trust. We shall show how the insertion of money and accounting ledgers can equip all agents with a kind of memory.

Our starting point in Promise Theory is thus that no agent, whether material or immaterial, has an intrinsic trustworthiness or value, and so the 'value of value' is limited, as an assessment, because it is expensive and unreliable to make inferences about. Value is an assessment made by every agent individually, and this assessment may be informed by individual preferences along side shared social conventions, and is relative to the context or circumstances of the agent. Trust is only one such valuation of agents, but it is a universal one in the sense that it can act as a weak proxy for all other valuations.

According to Game Theory, so-called rational economic value is formed from the certainty of repeatable (or so-called 'invariant') characteristics, which persist over repeated interactions[Axe97, Axe84]. We shall return to the question of value more thoroughly in section 2.1. Although properties may change in the long run, their short term stability or 'invariance' is the key to their usefulness. This iterative revisitation of interactions between agents that behave predictably by keeping their promises is the basis of trust[BB06b]. One may construe that these invariant properties are intended for use, effectively promised by someone or something (by proxy[5]). Promises represent a labelling of intent, to offer or to make use of invariant characteristics[6].

Measuring and maintaining trust is thus expensive. It costs time and effort, thus it is natural that trust would be linked to a notion of value[BB06b].

Example 2. *Credit scores have become a modern day surrogate for monetary trust. One could even pose the question whether trust is a currency whose proxy is a credit score. Credit-worthiness, in the eyes of some Trusted Third Parties (trusted by banks) are now a de facto measure of societal trust.*

Example 3. *Technologists often misuse the term trust to imply its complement* verification. *In information technology, trust is often assumed to be a state in which one has verified a fact, rather than having avoided the need to verify a fact. This leads to some mixups in semantics. For instance, the use of blockchain technologies to validate communications allows exterior systemic trust through low level interior validation.*

Trust is an exterior assessment of a single agent (or superagent), whose interior details we don't want to know on a regular basis.

2.1 VALUE AND UTILITY

The concept of money has long been closely entwined with that of *value*, or *utility* in the history and literature of economics. We need to separate these concepts. Common usage has two origins[Mir89]:

- The belief that agents might be guided by a kind of 'preference potential'. Since Jeremy Bentham's concept of a principle of greatest happiness to rationalize human behaviour, many social sciences have postulated an immeasurable phantom field called the 'utility', analogous to a 'potential energy' function in physics. In economics, utility is used to refer to the total expected satisfaction received from consuming a good or service. It is considered to be a representation of preference.

- The belief that goods might carry with them an intrinsic value, determined by strict accounting principles. This understanding of value has been muddled with monetary measure for much of the history of economic theory. An absolute measure of value seems to fly in the face of scientific experience however. It is much easier to claim 'I prefer A to B than 'A is better than B', although the proximate intent might be the same.

Over time, these concepts have merged into a single concept of *utility* that is assumed to guide their economic behaviour in a more or less deterministic manner. The work of Axelrod on cooperation and utility has informed this discussion[Axe97]. In both these cases, utility is regarded as a measure which 'rational agents' seek to maximize in making a decision. The discussion in [NM44] is perhaps the most careful summary of this.

As a behavioural indicator, utility is frequently assumed to play an immediate and causal (deterministic) role in economics[7]. Rational consumers are supposed to be directly motivated by their expected utility. It is even assumed to influence the demand for goods, and therefore their price. We cannot sanction these assumption here: the weight of evidence contradicting these hypotheses is considerable[Kah11, Ari12, Lay06]. We shall, however, mention some satisfactory aspects of utility theory worth preserving.

A definition of utility was carefully formalized by Von Neumann and Morgenstern in their treatise on economics games[NM44]. As a singular indicator of preference, they showed that utility could be represented by a scalar quantity U, which could be assessed in various ways, but would have to satisfy the following axioms in order to preserve preferences:

$$
\begin{aligned}
U_1 &> U_2 \\
\alpha(U_1) &> \alpha(U_2) \\
\alpha(c_1 U_1 + c_2 U_2) &= c_1 \alpha(U_1) + c_2 \alpha(U_2)
\end{aligned}
\tag{2.1}
$$

Here we follow the notation of [BB14a] in taking $\alpha()$ to mean an assessment function. Von Neumann tried to make this rigorous[NM44], but his argument was detached from worldly concerns. Game Theory treats utility as a temporary social convention, whose variation happens only slowly compared to the rounds of a game. Axelrod, Hamilton and the evolutionary biologists made much progress in defining utility as extended rounds of gaming (see [Axe97] and summary in [Bur13b]). Promise Theory recognizes any number of valuations in [BB14a] (more details were described in [BF05, BF06]). These are purely observational assessments, and cannot be directly causal without specific promises to that effect. We write:

- $v_i(\pi)$ as agent A_i's valuation (in its own units) of a promise π, by virtue of the existence of the promise. A valuation returns a number in some set of units, which must be promised. It could also return a tuple, indicating assessments along a number of independent criteria.

- $\alpha_j(\pi, t_i, t_f)$ is agent A_j's assessment of the extent to which a promise has been kept, assessed over an interval of time from t_i to t_f.

- $v_i(\alpha_j(\pi))$ as agent A_i's valuation (in its own units) of the assessed state of keeping a promise π, by virtue of the existence of the promise.

- Promises may persist, but an assessment is a transaction over a specified interval of time. Thus every agent's concept of value is transitory.

- Trust is a low specificity assessment, which is therefore fungible or applicable universally, in the same way that money is fungible and applicable almost universally. There is thus a natural relationship between trust and money[Har00, Bur17a].

Notice that, in all these expressions, each agent is not only free to make its own assessment; it is required to do so. If there is any consensus between agents, then this must be demonstrated or engineered by cooperation between them. We must not assume consistency, a priori; thus there can be no golden standard of valuation or utility. Such agreement requires strong conditions of social calibration and coordination. In fact, a simpler viewpoint is that money serves to replace these bungled assumptions with an invariant standard measure, whose relationship to value is entirely arbitrary.

Axiom 1 (Utility and use-promises). *The value of any object, good or service lies in an agent's belief in its ability to utilize the promises made by it at some later time. Thus utility is a valuation of a use-promise (see 3.10.2 and 9.1 in [BB14a]).*

Definition 2 (Utility). *A utility function $U(\pi)$ is an assessment of the usefulness incurred by accepting the offer in a promise with body b:*

$$\pi_+ \quad : \quad A \xrightarrow{+b} A' \tag{2.2}$$

$$\pi_- \quad : \quad A' \xrightarrow{-b} A \tag{2.3}$$

$$U(\pi_-) \quad = \quad \alpha_U(\pi_-) \tag{2.4}$$

This indicates that the utility of a promise is measured by the user (i.e. the acceptance of a promise given by π_-). It is a function mapping a promise body to a number satisfying the criteria in the axioms (2.1).

Utility is a valuation (i.e. a form of promise assessment) which being made by a human or its proxy may belong quite generally to the real numbers[8]. Like all assessments, utility is a 'belief' or expectation function, like a Bayesian probability. Money is an approximate discrete quantized representation of measure.

Axiom 2 (Money should be assumed approximate). *Although it is theoretically possible to maintain high (perhaps even perfect accuracy) in accounting, it is prohibitively expensive to do so, and never done (to our knowledge). Current forms of accounting represent only a finite number of decimal places.*

Rounding errors (deliberate or limits of technology) make money a discrete representation in practice. Usually, decimal places are not retained by banks and accountants. Thus money maps intended utility values into discrete (finite accuracy) units units that acts as a proxy:

$$\mathcal{M} : \mathcal{U} \to \mathbb{R} \otimes \mathbb{Z} \tag{2.5}$$

Not all currency values may be mapped by (isomorphic) scaling transformations of one another. We call currencies that may be mapped *congruent*. In fact, currencies are never really compared, they are bought and sold in markets. So transformations from one to the other take place by sale rather than accurate bijection. Thus a certain amount of intended money can be expected to go lost each year. This is a form of entropy (see section 6.11).

2.2   THE AXIOMATIC VIEW OF MONEY IS UNRESOLVED

The most basic mathematical question one would ask about money is: what is the algebra of money? In daily operations, we perform arithmetic operations of rings and fields on monetary amounts, but the finite accuracy alluded to above implies that money cannot be a ring or a field without a continuous policy of approximation[Gre88].

The associative and distributive properties are normally taken for granted in arithmetic. If μ_i for $i = 1, 2, 3, \ldots$ are monetary amounts, then it is easy to accept some of the axioms of arithmetic. Associativity would be the key to money's fungibility and loss of memory:

$$(\mu_1 + \mu_2) + \mu_3 = \mu_1 + (\mu_2 + \mu_3), \tag{2.6}$$

however, we need to know when rounding takes place in the addition of amounts. Is it after each addition from left to right, or when brackets are closed? Each policy requires some memory associated with a computation, and the deliberate loss of information on rounding. We might also not be able to assume the distributive property, for the same reason:

$$2(\mu_1 + \mu_2) = 2\mu_1 + 2\mu_2, \tag{2.7}$$

e.g. $\mu_1 = 2.3 \simeq 2$, $\mu_2 = 3.4 \simeq 3$, so that $2\mu_1 = 4.6 \simeq 5$, $2\mu_2 = 6.8 \simeq 7$, and $\mu_1 + \mu_2 = 5.7 \simeq 6$, and $2(\mu_1 + \mu_2) = 11.4 \simeq 11, 12?$, while $2\mu_1 + 2\mu_2 = 11.4 \simeq 11, 12?$. Should the result be 11 or 12? The order in which we round numbers matters. Rounding is a non-commutative operation. We shall not address this further, but consign it for future study. Rounding implies a growing uncertainty in the amount of money on a ledger. If all amounts were simple transactions of fixed monetary amounts this would be less of a problem, but we frequently want to rescale amounts by real numbers, e.g. 17% VAT or 15% off! This makes discrete monetary amounts ultimately inconsistent. The accrued loss of small amounts of money is a loss in the money supply, which likely goes unaccounted for. The loss is potentially borne by the receiver in a transaction. This loss of money may lead to a change in the relative effectiveness of the money in circulation, but the possible consequences of this are not clear to us.

2.3 NETWORK OUTPUT OR 'VALUE'

The economic output of a network can be measured in terms of its agents and its promises. Metcalfe's law claims that economic output of a network should be proportional to N^2 in the number of agents. This has been criticized theoretically (for a sample see [OT06, Bri06]). However, recently this conjecture has received empirical support from social media studies[ZLX15]. Metcalfe originally assumed that value creation would be proportional to N^2 while costs grew proportional to N. The study [ZLX15] indicates that both grow quadratically with network vertex count, though these ideas are still disputed[OT06, Bri06]. However, studies of the economic output of cities also vindicate this idea[Bet13, BLH$^+$07, Bur16a]. If value is proportional to the number of agents (e.g. workforce) then output is proportional to N. This is true of promises made by each agent individually, working in isolation. For economic output to exceed this, there must be a network component, as shown in the data of cities and their economies of scale. The universal argument has been given by Bettencourt[Bet13], and verified with empirical results for Gross Domestic Product (GDP) and a variety of economic indicators, and explained microscopically in terms of Promise Theory by Burgess[Bur16a].

Neoclassical economic theory is discussed with energy as an exogenous factor. In other words, the exchange of energy (which is a prerequisite for all processes considered valuable or value creating) assume that energy is paid for and supplied in a completely 'out of band' side channel. However, work on the scaling of biological organisms [Wes01] and cities [Bet13] in relation to the output and consumption show that the basic infrastructure, which communicates these prerequisites like energy and information, can explain the economic scaling of these networks at a coarse level quite well. In other words, we can say that economic output at scale is determined in large measure by communications, not by work or other internal factors.

2.4   ACCOUNTING AND TIME

Expectations about money have a time limit. For example, the idea that an economic system might reach an equilibrium is meaningless without a timescale. Such an invariant state may only emerge in a limit of very long times (too long to have any significance to any single human). On the other hand, the accumulation of 'payoff', in a game theoretical sense[NM44, Axe97], is iterative, like the ticking of a clock. Time passes every time money is exchanged[Hic73].

The allowance of a social concept of debt grants agents permission to exceed their immediate means and overcome hindrances, putting off repayment until some retarded date, or saving up in advance of payment. Retarded and advanced conditions of payment play an important role in the propagation of money, just as they do in signal transmission. This is further evidence of the role of money as a network communications mechanism. Moreover, the ability for banks to create money without delay and mediate between agents, by displacing the need for direct peer to peer trust[Har00], in favour of trusted institutions has accelerated economic activity and made it possible for societies to avoid much violence in exploiting personal concerns in repayment of debts. Trusted Third Parties, in modern language of information systems, such as lenders, banks and governments, eliminate much of the cost of having to build up a trust in individuals directly, bypassing individual relationships in favour of the impersonal state, and thus allowing individuals to believe that they could act safely with a lower risk of deception. Money asserts influence with *spacetime locality* by allowing parties to insert time delays for verification, yet still transact at a single proximate point of time and space.

Mathematical accounts of such 'exchange' and 'influence' have been developed in many different fields of study, but the most broadly developed of these is surely in physics, where the concept of 'energy' emerged over several centuries as a means of explaining the transmission of cause and effect[Mir89][9]. We shall not attempt to summarize this story here, but merely note that, in order to function successfully, energy is used as a bookkeeping quantity which tracks the movement of influence around a system. In order to fulfill this function, energy must be assumed immutable. Energy can move from place to place, and even be transformed into different manifestations, but it may not randomly disappear or be created spontaneously. This is an essential feature of a quantitative delivery system, somewhat analogous to the post office promising not to lose mail, but also one that gets harder to maintain as indeterminism creeps into system descriptions.

The post office analogy may be more significant than one might imagine, as there is a point of view in which information is the fundamental conserved quantity, and energy is only a proxy 'data packet representation' for this information. Ultimately, energy conservation is also a hypothesis, which cannot be proven without assumptions that merely shift the blame around. Yet, what is important is that it is a consistent point of view that leads to quantitative consistency of accounting. It is not hard to see that, to make sense of economic accounting one is also motivated to preserve this view.

Because the creation of money is independent of the creation of goods and services, and fluctuates in its influence in a non-conserved way, there is only one rational role for monetary conservation: the proper accounting of *trust*. Evidence for this comes from studies of the iterative games in Axelrod[Axe97] (see the discussion in [Bur13b]).

For this reason, we need to understand the role of time in economic behaviour, because we can only say that a promise to deliver or pay has been kept after a certain time has elapsed. If agents have to wait forever, their trust will be stretched beyond its limit.

2.5 CONSISTENCY AND ITS HOMOGENEITY

In the twentieth century Emmy Noether became known for her proof of physical conservation laws, based on the uniform continuity of spacetime. Translational invariance leads to conservation of momentum. Time translation invariance leads to conversation of energy, and so on. There are many other possible laws of conservation, but these all amount to the following assumption. Irrespective of whether spacetime is continuous, if it is homogenous and uniform in its treatment of the dynamical quantities (i.e. if there are no preferred locations or times that account for physical interactions differently) then one can show that what goes in must come out equally at all locations and times, meaning that nothing can go amiss. In fact, one does not need the continuum approximation to make the same argument for discrete agents (at any scale). Discrete time invariance, or conservation, is expressed by the Markov property, which also expresses memoryless propagation of a random variable.

Definition 3 (Markov process). *Let X_t be a random process, for totally ordered times $t = t_1, t_2, \ldots, t_n$[CT91]. A Markov property makes the sequential conditional promise*

$$\Pr(X_{t+1} = x | X_t, X_{t-1}, \ldots) = \Pr(X_{t+1} = x | X_t). \quad (2.8)$$

i.e. the next transition depends only on the current state of the process, not on any memory of previous states. A Markov chain or process is characterized by a transition matrix (the weighted adjacency matrix of the directed transition graph):

$$P_{ij} = \Pr(X_{n+1} = j | X_n = i), \quad (2.9)$$

which renders is equivalent to a non-deterministic finite state machine. The chain is said to be homogeneous or translationally invariant *if*

$$\Pr(X_{t+1} = j | X_t = i) = \Pr(X_1 = j | X_0 = i), \quad (2.10)$$

which implies the conservation of distributed expectation values, e.g. energy E_i

$$\langle E \rangle = \frac{1}{T} \sum_{t=0}^{T} \Pr(X_{t+1} = j | X_t = i) E_i = \Pr(X_1 | X_0) E, \quad (2.11)$$

which is independent of time t, T. The loss of memory in an evolving distribution, as time increases, is equivalent to the second law of thermodynamics $\partial H(X_{t+1} | X_t) / \partial t > 0$, where $H(a|b)$ is the conditional entropy[CT91].

Economic agents are not Markov processes as a general rule: they remember previous states, through accounting ledgers, and even in simplistic ways like 'trust', which has a Bayesian statistical character. So we should not expect time homogeneity in economics, without strong constraints. However, herein lies the problem for money: the agents in the spacetime of economic activity are very far from homogeneous and uniform. The treatment of money, goods, and value may be quite different, leaving no basis on which to argue that money has to be conserved. Indeed, we know that it cannot be automatically conserved by any reasonable law of nature or Man. Anyone can burn paper money, and crush coinage for jewelry with impunity, and that money is simply lost, with uncertain consequences.

By contrast with physical theories, the tendency for agents to interact (their 'coupling strength') is not a constant, but is 'dressed' by a changing level of *trust*[BB06b, BB14a]. Conservation of trust over multiple iterations is not guaranteed, any more than there is conservation of value or the interpretation of 'utility' between different contexts in Game Theory. Trust is a very simple form of Bayesian learning[Pea88] (memory) about past behaviour, as are utility and value.

What then might be conserved about money? Arguments have been made for all of the following: the material, information, value, energy, and work that go into representing it. We shall argue that what is conserved is a language of communication, i.e. the semantics of money, and that this is combined with a desire to perform strict accounting of amounts. The variations can then be renormalized into the setting of prices, which lie at the endpoints of an economic network.

2.6 CALIBRATION AND TRUSTED AUTHORITY (CENTRALIZATION)

A result of Promise Theory, which will play an important role in economic networks concerns the relationship between centralization and consistency. Centralization has the effect of *calibration*, which is sometimes interpreted as the semantics of 'authority'. Trusted 3rd parties allow us to remove the cost of verification between all pairs, of agents, and replace it by a much cheaper centralization (depersonalization) of trust in a *monetary authority*, with a significant reputation.

> **Definition 4** (Calibration of promises). *An aggregate assessment of multiple promises, in which all promises are received and given in the same context and under the same conditions.*

> **Lemma 1** (Calibration by centralization). *It may be shown that a calibrating agent implies* localization *or 'centralization' of assessments, which means the agent has access to promised information without delay or other impediment.*

This follows by (**Ax2**) and (**Ax4**). The role of decentralization and centralization of information routing is illustrated in figure 2.1. This figure illustrates the alternative ways of preserving information, and it recurs in multiple contexts in understanding networks. The accounting of trust is subtle. In scaling terms[10], we may not be able to say exactly how many agents are involved in an interaction, but what we can say is how the cost of maintaining a type of interaction will grow with the total number of agents N. In a centralized configuration, N agents connect with a single intermediary (e.g. the bank or software system). As the number of agents grows, this interaction grows in proportion to N, and there is a single calibrating agent, as in the right hand side of the figure. In a decentralized or cluster configuration, involving some fraction of N, each agent has to deal with each other agent, so that the growth in cost is proportional to some fraction of N^2, but not usually the full N^2.

For small N, the scaling cost of peer clusters is not important; indeed, this is reflected in the algorithms of computer science concerning consistent information and its

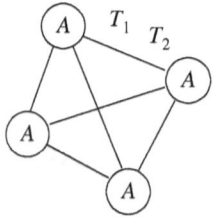
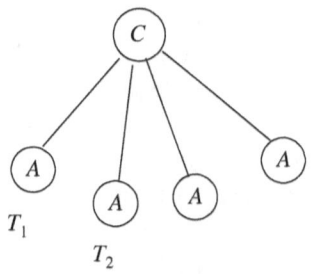

Free equilibration   Modular trust
Individual untrusted calibration   Central trusted calibration

Figure 2.1: In a network of interactions, centralization allows a (semantic) calibration through a third party, which is cheap $O(1)$ for everyone except the central entity, who has to deal with $\propto O(N)$ connections. In an equilibration, each agent has to manage $\propto O(N^2)$ interactions, so for $N > 2$ centralization has collaborative cost benefits for most, but may become a (dynamical) bottleneck for the central calibrator. This figure will recur many times in considering network dynamics and semantics.

relativity[Lam01, OO14, Lam78]. However, as N grows large, at constant processing time, it renders cluster processes quadratically slower than centralized ones. Even the distributed algorithms in computer science try to avoid this by centralizing processing, electing a single trusted 'leader' for efficiency.

Of course, with such an important role, any central agent (trusted third party) needs to be trusted beyond reproach. Today, governments guarantee banks to some extent, and interact with them to help them with financial stability, so trust in the third party is bolstered by trust in the government, usually by expectation of fair judiciary.

Lemma 2 (Distinguishable by O). *Let O be any agent in the role of observer, and let A_1, A_2 be any agents. Two promises π_1 and π_2 may be called distinguishable by O if and only if O is in scope of the promises and is able to accept the distinction:*

$$\pi_1 : A_1 \xrightarrow[O,\dots]{+b_1} R_1 \tag{2.12}$$

$$\pi_2 : A_2 \xrightarrow[O,\dots]{+b_2} R_2 \tag{2.13}$$

$$O \xrightarrow{-b_1} A_1 \tag{2.14}$$

$$O \xrightarrow{-b_2} A_2 \tag{2.15}$$

where R_1, R_2 are any agents and $b_1 \neq b_2$.

This follows by (**Ax2**) and (**Ax4**).

Distinguishability brings up an interesting question about the role of time in promises. The moment at which a promise is assessed may be considered to play a role in comparing outcomes. If so, this has to be included as part of the information in the promise body, else there is no way to make the distinction between different times[Bur19a]. The role of time in economic transactions is subtle because, as is true for all distributed systems, there is no natural clock by which all transactions may be calibrated. We refer readers to [Bur19a, Bur19b] for a discussion of these subtleties.

2.7 PREREQUISITE DEPENDENCIES OF ECONOMIC AGENTS

The principle of autonomy is helpful in that it leads to a separation of concerns, and it is a reasonable model of political and national interests. It forces us to declare all hidden dependencies, if accounted for properly. Any agent that does not acquire something from another is assumed to contain that resource within. Energy is one dependency that no agent can have internally in infinite supply. Thus, for correctness, the supply of energy and other constituents, like raw materials and time, should be included in the accounting (but usually isn't).

CHAPTER 3

GOODS AND SERVICES

Goods are ownable agents; they may be bought and sold in an economy, along side services (which cannot be owned) and investments that are something in between. This does not mean they have to be physical items: they might be ideas, intellectual property rights, companies, virtual shares in some enterprise, etc. Agents have no properties *a priori*; only their promises distinguish them: thus what distinguishes a good is one or more promises. Thinking more carefully about the structure of goods and service as agents that make promises will give us some practice at thinking in terms of agent clusters. This kind of semantic scaling is at the heart of economic systems.

Goods are treated in a variety of ways in the economic literature, but most often they are handled as a kind of universal substance without distinct identity. Whether, in the final analysis, it is possible to disregard semantic distinctions between different goods we shall not take a position on at this stage. However, if it is indeed the case, we would expect this to emerge from an analysis at the microscopic level, through scaling, as in [Bur16a]. Here, our goal is to explain the economic theory of money and its function in buying and selling, thus we must retain functional qualities at least initially.

3.1 GOODS DEFINED

We tend to think of goods prosaically as physical 'items', but this is not an accurate representation of most goods. What we buy, as a good, is effectively a *concept*, which appeals to its potential buyer: the promise of a solution (often called a 'pain pill' or a 'vitamin' in contemporary business marketing). In many cases, it is not the physical reification of something we care about (except perhaps in the case of raw materials); rather, it is the concept that was forged from the physical form, or more precisely the promises a thing can keep to us as consumers (+ promises). The value of such promises lies in what use a buyer can make of them, i.e. in their level of acceptance (- promises). The promises of goods might be functional descriptions, like technical specifications, but they could also be branding labels and lifestyle claims, in the case of fashion and diet products. No one could deny the importance of promises to the marketing of goods in the modern world.

Example 4. *The physical reification of a concept may have several alternatives, some of which may be in competition. For instance, 'shoes', 'energy drinks', 'tea', etc are concepts that can be realized in multiple ways.*

Definition 5 (Good(s)). *A bundle of promises Π_{good} made by a cluster of one or more agents (i.e. made by the superagent formed by the coherent good) that are sold as a unit. A good promises a representation as a holdable thing T, whose value is assessed by the promisee (-), and is offered by the promiser (+).*

The natural association here is that the promiser is a seller and that the promisee is a buyer, but we can avoid those terms until we have defined them later in section 7.2. In the case of continuum commodities (liquid, powder, by weight or volume), the agent promises the transactional amount of those. Agents may or may not be combined into single equivalent agents (like barrels of oil).

Lemma 3 (Goods are subjective). *It follows from the definition above that the perception of goods is dependent on both promiser and promisee, since both (+) and (-) promises are involved in the communication.*

Users may or may not accept all the exterior promises of a good on an individual basis. Indeed, a buyer is free to probe the details of the parts separately. The buyer might accept the physical promises (e.g. tastes like chicken), but not accept or believe the branding promises (e.g. best in the business). Clearly, it is not the physicality of goods that is key to its economic usefulness within a network.

We define a good G_i to be an agent cluster, or super agent, which may be composed of several subagents. The entire cluster makes the promises of the good. Some are shared with other instances, others are unique such as the particular physical instance, with serial number and paint colour, etc. For example, the price of a good may change even if the price of its packaging changes, as this is an integral part of what is sold as a unit.

3.2 WHAT PROMISES DOES A GOOD MAKE?

While being strictly voluntary, our common understanding of a good is that of a cluster of agents that collectively promise the following:

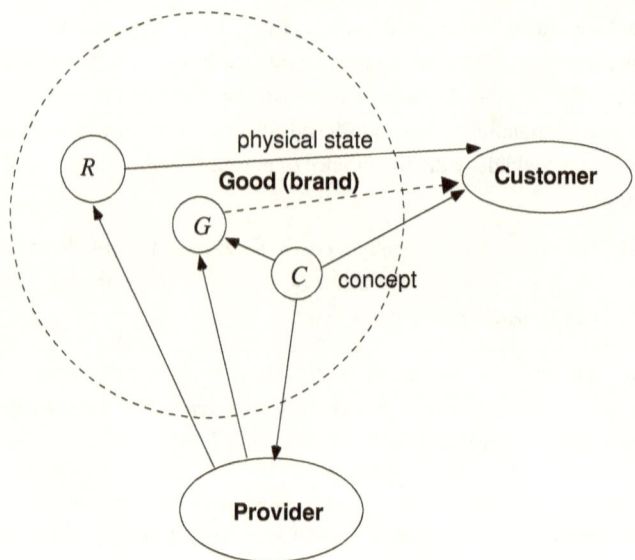

Figure 3.1: A good is a cluster of agents, some physical, some conceptual or virtual. R represents a realization or reification of the concept, and a unique instance, while C represents conceptual promises, brands and the labelling, packaging, etc. These sub-agents within the cluster belong to different spaces. Both G and C are only name for ideas: C is generic idea, and G is a specific instance.

1. The ability to be owned (and hence be bought and sold).

2. Promises to originate from a provider/manufacturer.

3. Promises to represent (be the reification of) a concept (originated by the provider).

There are two interpretations:

1. The good imposes a concept on the consumer, or

2. The good promises to represent a concept.

In practice goods are imposed without any dialogue, but services may be collaborative enough to support negotiated conditionals.

    At what point do these decisions become unimportant for the larger scale movements of an economy?

Example 5. *The promises non-consumable goods make will undoubtedly change of the lifetime of each particular instance: houses get remodelled, renovated, demolished, etc, and the value of the good may fluctuate in correspondence over its lifetime, appreciating*

product promises

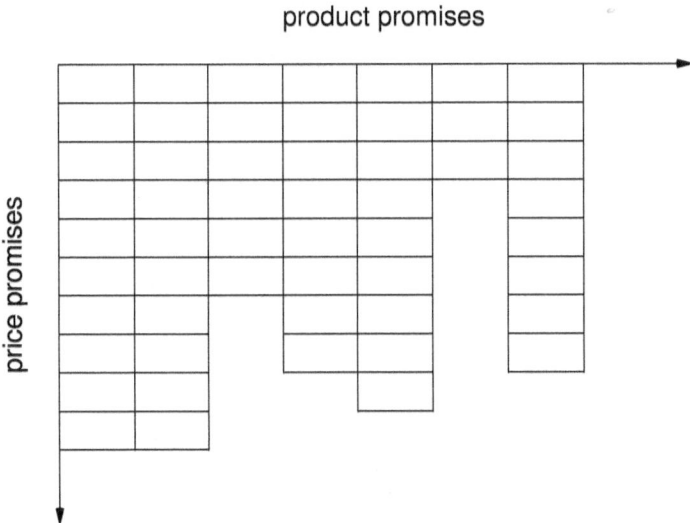

Figure 3.2: The periodic table in chemistry is a kind of histogram of semantic classes. The space of products (goods and services) and their partial ordering by size or expense is analogous to the periodic table of elements in chemistry. In each column there are products that make similar promises, and in each row, there are products whose price is about the same size. The analogy ends there, however.

and depreciating accordingly. A certain good may even disappear, be disintegrated and recycled, and yet it remains economically active, being mentioned in other promises, rumours, reputations, as well as in contracts, debt acknowledgments etc. A particular good may never be repaid even after it has been destroyed.

Assumption 1 (Value of goods). *The value of a good is a promise assessment and should be treated as a random sample variable. It changes over time and is context dependent.*

The value of a good has a finite lifetime both because a promise can degrade over a timescale Δt_{perish} and an agent may be fickle and change its mind on a timescale $\Delta t_{\text{preference}}$.

3.3 SERVICES DEFINED

Services are simply promises to exchange behaviour for payment. The result of a service is an outcome rather than a physical item (though the delivery of a physical item might be the outcome).

Definition 6 (Service(s)). *The promise to deliver an outcome $+o \in S$, by a service provider agent S to a recipient R, one or more times.*

$$S \xrightarrow{+o} R. \tag{3.1}$$

The service promise is kept if the full specification of the outcome is met (including any repetition of service, etc).

If o belongs to the set of services S it is a service, by definition, so there is a tautological aspect to the idea of a service. Services are associated with cooperation: one agent acts on behalf of another, thus if goods refer to nouns, the services refer to verbs.

Example 6. *Transportation, network delivery, money transfer, money lending, labour, work, delivery, arrangement, sorting, maintenance, are all examples of services. The outcome of delivering or handing over a good is formally a service, but perhaps a trivial one in proximal trading.*

3.4   PRODUCTS AND THINGS DEFINED

It is cumbersome to continually specify the distinction between goods, services. The marketing term 'product' is quite useful to remove the distinction between goods and services. We use of here when it would be cumbersome to say 'good or service'.

Definition 7 (Product). *A neutral alias for a good or a service.*

However, it seems unnatural to describe a used car as a product. The term such as consumable seems equally cumbersome. We shall therefore coopt the term 'things';

Definition 8 (Things). *Refers to any of the following: goods $\mathcal{G}$, services $\mathcal{S}$, monetary amounts $\mathcal{M}$ (possibly in other currencies).*

$$\mathcal{T} = \{\mathcal{G}, \mathcal{S}, \mathcal{M}\} \tag{3.2}$$

We use indices T_a where a runs over the members of a set of things to refer to individual buyable items in a set.

3.5   ASSETS, LIABILITIES, AND INTERBANK EXCHANGE

The concept of assets has a common meaning and a technical meaning in finance. The essence of both is captured by the following:

Definition 9 (Asset). *Any beneficial collection of promises* Π_{asset} *received by an agent A an agent, that is assessed, by an observer O, as having positive value to its owner* $v_{O \to A}(\Pi_{asset}(A)) > 0$ *through goods or service relationships, by any observer, e.g.*

$$\Pi_{asset} : \text{Thing} \xrightarrow{+\text{attributes}} A_{\text{Recipient}} \qquad (3.3)$$

$$v_{O \to A}(\Pi_{asset}(A)) \quad > \quad 0. \qquad (3.4)$$

Notice that an asset is measured by its promise, not necessarily by its being a physical agent. This is important, because many financial assets (in the technical meaning) are simply ledger entries for things that do not have any physical form. For example, a deed of ownership might be called an asset. We are most familiar with property deeds, but deeds can also be issued for nothing at all. Indeed (pun intended), such financial assets are the way that money is created and destroyed in an economy (see section 5.17). Notice, also, that time is ambiguous here: we do not say whether the assessment of value refers to the past, present, or possible futures. The same is true of the converse:

Definition 10 (Liability). *Any non-beneficial collection of promises* Π_{asset} *received by an agent A an agent, that is assessed, by an observer O, as having negative value to its owner* $v_{O \to A}(\Pi_{asset}(A)) < 0$ *through goods or service relationships, by any observer, e.g.*

$$\Pi_{asset} : \text{Thing} \xrightarrow{+\text{attributes}} A_{\text{Recipient}} \qquad (3.5)$$

$$v_{O \to A}(\Pi_{asset}(A)) \quad < \quad 0. \qquad (3.6)$$

Although 'asset' and 'liability' characteristics of a set of promises, which must originate in an agent (e.g. a good or service provider), it is an assessment which drives this role, and is thus a relativistic assessment.

If the observer chooses not to keep its assessment to itself, the characterization could be promised:

$$O \xrightarrow{+(G \text{ is an asset to } A)} \text{Someone.} \qquad (3.7)$$

However, this is a 'promise of the second kind'[BB14a], which violates the autonomy of G and A, and so its status is mere hearsay. When we talk about assets and liabilities, we understand them to be an alias for promises made by goods or services to the holder or owner A, whose status is a subject of speculation.

In the communication of money between banks, so-called 'financial assets' represent a kind of currency for exchange. Since money of account cannot leave the bank that created it, without losing control of its proper accounting, bank exchanges occur by selling 'assets' to one another.

> **Definition 11** (Financial asset). *A contractual claim or 'IOU' sold by a bank, in some currency, in order to create money in its own currency for an external or foreign buyer.*

Many so-called assets are not real things, just information. When a bank buys an asset from another, it does not receive a fabulous home by the sea, just a contractual claim, or a 'bond'. These can be sold for money of account, to be returned at a later time, with interest. This is used for borrowing between banks. It is, after all, simpler to invent a fictitious and reusable kind of artefact than to negotiate exchanges on a case by case basis. As we shall see, later when discussing markets, it is a common strategy to commoditize things and thereby render their information compressible and cheap to . They can also be used to legitimize foreign exchanges between different currencies, being easier to sell than foreign property or real assets.

3.6   THE CREATION AND DESTRUCTION OF THINGS

Things can be created and destroyed, though this is not directly relevant to the story of money. In the classical treatment of economics, e.g. by Marx, the origin of value added to raw materials is due to the combinatoric creation of new things from the raw parts. Thus integration and disintegration could be part of the creation and destruction of 'value'. Yet, when things come and go, money remains independently of these. In case it still needs to be pointed out to readers, neither discovering gold, silver, growing fruit, nor catching fish lead to the creation of money. These things cannot be sold for money unless sufficient money exists independently. The point of interest here is that there has to be sufficient 'liquid' money available to enable the sale of these new things.

In figure 3.3, we illustrate the creation and destruction of money. For completeness, we note that goods and services refer to concepts, not only to physical proxies, so the destruction of the proxy does not necessarily destroy the concept of the good, but it eliminates its promise of availability so that it can no longer be held. The promise of a good can only be destroyed by destroying all the physical proxies and removing the brand promise. Similarly, loss of goods is different from their destructions. For services, the instigation or removal of the promise to carry out the service is sufficient to do this, since a service is by definition an intent to act.

Customers create things, which neither creates nor destroys money. Money must exist already to be able to buy and sell new or old things. Banks can create (+) and destroy (-) money by accounting. The central bank can create money autonomously, but other banks B_i can only create money conditionally on the state of their balance sheet (ledger). Customers can also destroy money by destroying its proxies, but cannot create authorized money. In the case of cryptocurrencies, the authorized software algorithms

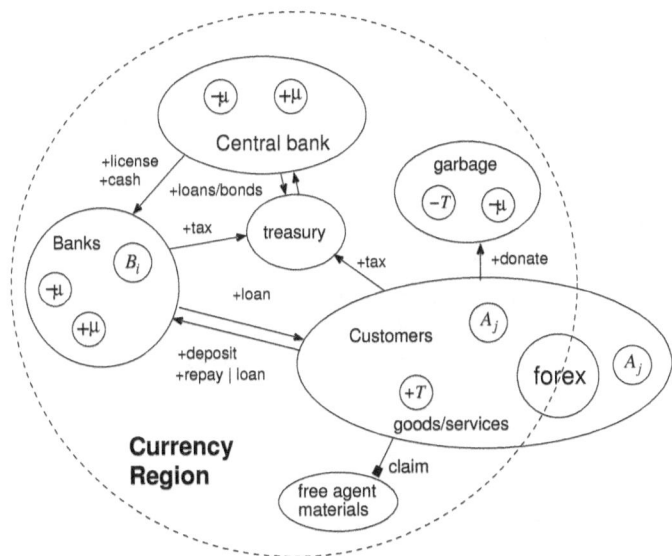

Figure 3.3: The scale at which authorized money and things are created and destroyed.

represent the central bank authority to join the system, and all customers are 'bankers'. Foreign exchanges can only buy and sell currency. Customers can exchange currency for things, but cannot create money. Ultimately all money is therefore created by a licence to the central bank. Finally, we introduced a garbage agent, as a symbiotic resting place for all such destroyed or lost things. In this way, one can account for the things, and pretend their conservation, analogous to the conservation of energy. This is simply good bookkeeping practice.

3.7 LIQUIDITY, AVAILABILITY AND AFFINITY

Liquidity is a concept used freely in economics. According to Investopedia, it is:

> The degree to which an asset ... can be quickly bought or sold in the market without affecting the asset's price[Inv17a].

Cash is liquid, but property is not. Liquidity suggests a timescale: how quickly could something be sold (in current market conditions) to yield its measure as money? In other words, relative to a given set of circumstances, which are assumed to be unaffected by the desire to sell, liquidity refers to the ease of being able to sell in exchange for something else (which is unspecified). The condition that a sale should not affect the price, is not an achievable condition, since it could only apply to bulk commodities. The sale of unique items will always change price when sold, e.g. property. Thus the liquidity concept is preferentially leaning towards bulk sales, like a large thermodynamic reservoir of intentional activity. However, this seems unnecessarily restrictive, and we can do better than this without violating the laws of information.

In accordance with Promise Theory axioms, we can break down the definitions into two parts, by defining the availability and affinity of X within a network region.

Definition 12 (Network promise of availability of X). *Let S be an agent in possession of X, which may be any kind of thing. We define the availability of X to R, subject to a set of conditions c_i waived by agents A_i, over time interval Δt, by the necessary and sufficient bundle of promises:*

$$\Pi_{+X} \equiv \begin{cases} \pi_1^+ : S \xrightarrow{+X|c_i} R \\ \pi_2^+ : A_i \xrightarrow{+c_i} S \\ \pi_3^+ : S \xrightarrow{-c_i} A_i, \end{cases} \tag{3.8}$$

Lemma 4 (Assessment of availability). *Since all three promises are necessary, an observer's assessment of the availability of a thing is the product of the assessments $\alpha_O()$ of the independent promises being kept:*

$$\alpha_{available} = \alpha_O\left(\Pi_{+X}\right) = \alpha_O(\pi_1^+)\alpha_O(\pi_2^+)\alpha_O(\pi_3^+). \tag{3.9}$$

This follows from (**Ax3**), (**Ax4**), the treatment of assessments as probability-like measures, and the rules of independent probabilities, as does the following:

Definition 13 (Network affinity of a thing X). *Let R be an agent in possession of X, which may be a thing or monetary amount. We define the affinity for X by R, subject to a set of conditions c_j waived by agents A_j, over time interval Δt, by the necessary and sufficient bundle of promises:*

$$\Pi_{-X} \equiv \begin{cases} \pi_1^- : R \xrightarrow{-X|c_i} S \\ \pi_2^- : A_j \xrightarrow{+c_j} R \\ \pi_3^- : R \xrightarrow{-c_j} A_j. \end{cases} \tag{3.10}$$

by the product of the assessments $\alpha_O()$ of the independent promises being kept:

$$\alpha_{affinity} = \alpha_O\left(\Pi_{-X}\right) = \alpha_O(\pi_1^-)\alpha_O(\pi_2^-)\alpha_O(\pi_3^-). \tag{3.11}$$

We define these assessments $\alpha_?$ to be estimated on support from the region $S \cup R$ over Δt. From the Promise Theory axioms, these two compound promise measures are necessary and sufficient bundles, which assure conditionally the intent to transfer X from one (super)agent to another.

In order for something to change hands, there must be a promise to offer it by a seller, i.e. an abstract *availability*. There must also be a promise to accept by a recipient, i.e. an abstract *an affinity for the thing*. We shall use these abstractions in the following chapters to define markets.

In the physical sciences there is a concept of *mobility*, which measures how likely it is for a thing to change its holding position. The corresponding term used in economics is *liquidity*. The liquidity of a thing is not an intrinsic property but a mutual property of promiser and promisee. This is what we shall use to argue that markets are effectively information channels with seller and buyer in the roles of sender and receiver. They thus contribute to a network of such channels. Liquidity is therefore not a property of a thing or of money, but an assessment about a set of circumstances in which promises are offered and accepted. From the axioms, we offer the following natural interpretation of liquidity:

Definition 14 (Liquidity of X). *An assessment by an observer O of the probability of an exchange via a sale in which $X \in \{T, \mu\}$ passes from S to R, during a time interval Δt. Liquidity is the product of the independent availability and affinity for a thing, within a network region, over an interval of time Δt.*

$$\alpha_{liquidity} = \alpha_{available} \times \alpha_{affinity} \qquad (3.12)$$

Liquidity level is reduced by conditions and encumbrances on the promises to offer or accept.

Liquidity of X may refer to money proxies or to things exchanged. As a probability it can only refer to a specific spacetime interval, since it cannot be computed or estimated without making reference to such an interval. As in all cases, Promise Theory emphasizes that it is the receiver who makes the final constraint on both dynamics and semantics.

Further averages across markets and multiple sales events could be used to define *average liquidity*, which is the most likely interpretation of the concept used by economists. We note that, as an assessment, it cannot be made very accurately because it requires there to be a convolution of distributions, which are the results of independent or even random processes. So average liquidity attempts to measure the shape and overlap of two distributions. Liquidity should ideally be defined in terms of the generalization of promises to market channels (see section 8).

CHAPTER 4

OWNERSHIP, HOLDING, AND EXCHANGE

To make sense of money, we have to begin with a fundamental assumption: namely that things can be held and owned as property. The idea of trade, of buying and selling goods, as well as the justification for exacting rent on property or services, depends on a convention that has its roots in the notion of ownership. We therefore need the definitions of ownership as a prerequisite to discussing trade, goods, commodities, and even money. The concept of ownership is a semantic social convention, to be sure, but one without which buying and selling can have no meaning.

The desire to track the location and ownership of things is rooted in an idea from physics, known as the principle of *local conservation*. It plays a central role in understanding the accounting of physics, and a corresponding role in economic accounting. To keep proper accounting, we need to keep track of semantics too, just as physics uses a multitude of thermodynamics potentials to distinguish the changing embodiment of 'energy'. Without the concept of ownership (which may have to be defended against contesting claims) there would be no impediment to agents merely taking anything at random, in a disorderly manner. As in the rest of the biosphere, this would lead to conflicts of interest and state of ecosystemic instability.

Ownership represents a form of structural persistence that brings semantic stability to our dealings with other agents in a world that has few natural boundaries on the scale of individual actors[Bur13b]. Ownership is not only a human social convention, but can be found in the individual and collective territorialism of many species.

Assumption 2 (Ownership and property). *Agents may be property, and they may own* *property.*

Having observed ages of human history in which everything from materials, to animals, and even humans were claimed as property, this seems to be all we can say on this matter. No moral judgement is associated with this observation—indeed, Promise Theory tells us that each assertion of ownership is for agents to *assess* and potentially reject. Similarly, the ability for agents to own property has been assigned to humans, animals (pets are often assigned inheritances by their human owners), companies, countries, and even artificial intelligences[Woo19]. Ultimately, the ability to own must be a social convention, like the law, that works as long as a sufficient number of agents accepts it.

Ownership is related to the occupation of resources. Indeed, when invading forces take control of territory, they 'occupy it' to lay their claim, until other agents promise to accept their right to own it without actual occupancy. It will be important to distinguish between occupying, holding onto something, and actually owning it. These represent different kinds of promise bundles. Occupancy of space, in turn, is similar to 'being held'. It relates to the spacetime composition of agents (e.g. of one agent being within the property or estate of another), and its derivative notion of tenancy, were discussed in [Bur15], and we base this discussion on that one. In particular the idea that agent clusters can emit and absorb agents within them was introduced in section 3.5 of that citation, with only cursory detail that will be expanded on here.

Ownership is not the same as holding, however. One does not have to hold something to own it, though sometimes agents will hold on to something to claim it as their own (see section 4.7). Indeed, without this concept, money of account would be impossible, since the agents apportioned ownership of certain funds in bank money never hold it (it never leaves the bank). Similarly, if one loses something, without a registered claim of ownership (or the label is removed or degrades), it may effectively return to being in a free state. This bears a relationship to Shannon's symbolic error correction theorem here[SW49]. Data corruption destroys the relationship of information to a sender/receiver. One can no longer say that a symbol was intended for the receiver if it is no longer what the sender intended.

In this section, we explain further the detailed application of the autonomy rule to the question of ownership.

Example 7. *Ownership is a social concept, so it an assessment made by an observer of some mental sophistication. It is not conventional to think of oxygen as being owned by a water molecule, as a pair of shoes is owned by a person, but the structure of the molecule lends itself to this interpretation, and we may observe a similarity of structure in these two cases, for all relevant intents and purposes, without muddling in too human a notion of free will or intent.*

4.1 HOLDING OF AGENTS BY OTHER AGENTS (CONTAINMENT)

One agent may promise to be holding another (e.g. you might hold coins in your pocket, or money deposits in your bank account). Holding of an agent is absorption by an agent, in the language of [Bur15].

Holding is a constraint on the freedom of a thing T by a holding agent H. We can express it in the usual way as an assisted promise. Not all agents may be holdable.

Definition 15 (Holdable agent). *To be holdable, T must promise some attribute that H can use to constrain T (like a hook, handle, tether, electromagnetic charge etc):*

$$T \xrightarrow{\;+attribute\;} H. \tag{4.1}$$

The holder H can use (accept) this attribute to interface with T:

$$H \xrightarrow{\;-attribute\;} T, \tag{4.2}$$

and use it, conditionally, to tether or contain T:

$$H \xrightarrow{\;+contain \mid attribute\;} T, \tag{4.3}$$

assuming that the thing T accepts and responds to this constraint:

$$T \xrightarrow{\;-contain\;} H. \tag{4.4}$$

This comes from (**Ax2**). Thus we can describe 'holding' entirely within the framework of voluntary cooperation, illustrating how this obligation free description leads to no loss of generality in practice. The detailed mechanism of holding a particular agent thus depends on its properties, of both agents, through the meanings of the tethering 'attribute' and the 'contain' force.

Example 8. *One can try to tether another agent by providing an incentive for its voluntary capture, or by attempting to coerce the agent's behaviour.*

When agents hold others, they form clusters, which we call 'superagents'[Bur14]. The identification and naming of a cluster is a form of scaling, in the sense that a cluster of agents promises to remain affiliated under some common aegis, and thus takes on the appearance of a larger entity that acts as a single agent of larger size. Other agents may or may not be aware of the interior composition of the larger agent[Bur15]. They may treat it as a 'black box', or as a collection of smaller parts depending on the nature of their interactions. This is how organizations, companies, organisms, communities, and nations work. Any agent can extend its reach by associating with other agents to form a superagent cluster. We can then ignore the interior details and treat the superagent as a black box. The black box must then be able to emit and absorb other agents, adding or removing from an unseen interior cluster. This is the simple model we can use to express the scaling of agency, as well as ownership, buying, selling, exchanging, etc.

4.2 OWNERSHIP AS EXTENDED AGENCY AND BLACK BOXES

Ownership is a step beyond agents 'holding onto' other agents, and turns out to be quite a complicated matter. We shall try to explain it in stages. As we shall see, holding might initially be used as a step along the way to appropriating or stealing something. The ability of agents to combine into clusters or superagents, capable of making extended promises is key to building up complex semantic behaviours like ownership.

Absorption and emission of agents that reside within other agents are one way of describing 'holding' of something. A_{sender} absorbs A_{sub} as a unilateral transaction. In [Bur15], definitions 19 and 20 of section 3.5, this was represented by the simple statements:

- Emission of a body part:

$$A_{sender} \xrightarrow{\ +A_{sub}\ } A_{recipient} \tag{4.5}$$

$$A_{sender} \xrightarrow{\ A_{sender} \rightarrow \{A_{sender} - A_{sub}\}\ } A_{recipient} \tag{4.6}$$

- Absorption of a body part:

$$A_{recipient} \xrightarrow{\ -A_{sub}\ } A_{sender} \tag{4.7}$$

$$A_{recipient} \xrightarrow{\ A_{recipient} \rightarrow \{A_{recipient} + A_{sub}\}\ } A_{sender} \tag{4.8}$$

Here, is it worth detailing these interactions more fully to give them substance. These descriptions are schematic, and can be given more substance[11].

4.3 OWNERSHIP AS A SOCIAL CONVENTION

Ownership is a state that requires a shared understanding amongst more than one agent. An agent can evolve through various states or levels of ownership (see table 4.1):

In a Promise Theory picture, everything is an agent so both the owner and the owned are agents that make different kinds of promises. To fully understand this, we need to understand the chemistry of how atomic agents form 'molecular' clusters with more complex properties. Ownership is, after all, effectively about the clustering of things by

PROMISE	STATE	PROMISES / IMPOSITIONS
π_{free}	Free and ownable	Default state: owned by * or $\emptyset$, or self
		and receptive to a new owner
π_{contain}	Held	Proximity promise binding
$\not\pi_{\text{claimable}}$	Not ownable	Rejects an imposition to appropriate it
		or can't point to its owner
π_{claimed}	Claimed / appropriated	By exterior imposition
π_{owned}	Owned and tagged	By promise transfer

Table 4.1: Stages and levels of ownership promises.

some semantic bonding. We tend to assume that the 'right' to own something is a simple matter of law, but that brushes over many complexities with sweeping assumptions that we need to unpick.

Let's begin with what is probably the way most of us have come to think about property, in order to see how this carries with it many assumptions that violate basic principles of agency. There are three main ways we might try to own something:

1. *Appropriation (decentralized autonomous claim)*:

   An agent A makes the promise that it owns a thing T:

   $$A \xrightarrow{\ +\text{I own } T\ } *. \tag{4.9}$$

   This doesn't make a promise on behalf of T, but works around it by implying an unstated imposition on T,

   $$A \xrightarrow{\ +\text{I own you!}\ }\blacksquare\, T. \tag{4.10}$$

   Thus it is an attempt to induce cooperation of T or trample its autonomy. The agent T might nevertheless accept this:

   $$T \xrightarrow{\ -\text{I own you}\ } A. \tag{4.11}$$

   More sinisterly, other agents might accept the promise of ownership without T's acceptance:

   $$A' \xrightarrow{\ -\text{I own } T\ } A, \tag{4.12}$$

   which equally implies an implicit imposition on T:

   $$A' \xrightarrow{\ +\text{A owns you!}\ }\blacksquare\, T. \tag{4.13}$$

   Finally, we note that there is nothing to prevent multiple agents from making the same promises, leading to contention. This is the problem with impositions (see section 4.7).

2. *Ledger registration (centralized authority)*:

   A ledger approach, in which a registrar assigns ownership allows there to be a central authority whose function all agents accept to be the state of things. Since there is only a single ledger, it is easy to avoid dual ownership contests (see section 4.9).

3. *Voluntary cooperation*:

   The thing whose ownership is being discussed T can promise its owner by pointing to it (e.g. a name tag or certificate of purchase, etc), and the necessary promises to accept this labelling can be acquired. This builds on decentralized information rather than a central registry (see section 4.6).

The only consistent solution that respects the autonomy of all agents in the final one, but it cannot be applied to all agents, because not all agents are capable of making the necessary promises.

4.4 AUTONOMOUS DECLARATION OF OWNERSHIP

An agent may promise equivalently to be the property of the owner.

Definition 16 (Owned by or Property of A). *An agent T is owned by another agent A if a promise π_{owner} is made to belong to a particular agent or superagent owner, by itself or by a third party (container or registrar), and the agent is able to promise its own unique identity.*

$$\pi_{owned} : T \xrightarrow{owner=A} * \qquad (4.14)$$

This is the only plausible proof that may be offered about a concept like ownership, which is compatible with the autonomy of agents and which does not rely on third parties to hold the information. However, not all agents can be labelled in this way. The water in a river can't be labelled, even though the river channel can be.

An agent may be its own property (which may be assumed the default state of agents). The default state of agents that are intentionally manufactured by another agent could reasonably be assumed to be the parent agent. In other cases, it is more natural to assume that agents begin as free agents. These 'pure' definitions should be compared to the promise of a claim of indirect ownership by a third party registrar in section 4.9, where agents also need to promise distinguishability.

Ownership begins as a number of promises of interior intent. The concept of ownership can exist in each agent independently, but if each agent has a different idea

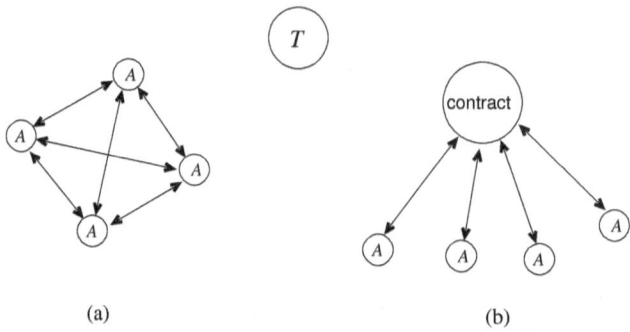

Figure 4.1: Ownership involves agreement about some social conventions. This can be agreed (a) directly between all agents on a peer to peer basis, or (b) by agreeing to a calibrated central standard. These two network structures are of fundamental importance in Promise Theory; they recur repeatedly as alternative ways of coordinating intent.

about the concept, this leads only to contention and uncertainty. We therefore assume that all agents, a priori, *agree* on the basic concepts of property and ownership as a mutually exclusive state of agents, either by a self-calibrated calibrated consensus (see figure 4.1):

$$\pi_{\text{peer conventions}} : \left\{ \begin{array}{l} \{A_i\} \xrightarrow{+\text{property rules}(T)} \{A_i\} \\ \{A_i\} \xrightarrow{-\text{property rules}(T)} \{A_i\} \end{array} \right. \tag{4.15}$$

or, alternatively, by a central calibration, let us call it a societal or 'legal' standard:

$$\pi_{\text{agreed conventions}} : \left\{ \begin{array}{l} \text{contract} \xrightarrow{+\text{property rules}(T)} \{A_i\} \\ \{A_i\} \xrightarrow{-\text{property rules}(T)} \text{contract} \end{array} \right. \tag{4.16}$$

This role assignment of a specialized agent is an example of the network pattern described in figure 2.1. The latter for in (4.16) is considerably cheaper to maintain (being of order N rather than N^2 for N agents of any scale). The property rules will include the idea that agents may be owned by named lists of agents, and that others who are not on the approved list do not own the item. All agents must recognize one another's names. One hopes these names are unique, but that is not something easily promised or imposed, without a central authority (such as a bank with an issued account number). Thus there is a legal convention or consensus basis for property and ownership, on which the remaining promises rest.

- Agents may lay claim to 'free agents' which they consider to be assets, discovered within the bounds of their own agent territory. These may, of course, be disputed and thus agreement about ownership is the basis for cooperation and 'international law'.

- Agents needs to agree on the definitions of their borders.

- Free agents that might be claimed as property may or may not have the capabilities to promise to be at a well-defined spacetime location. Cars and raw materials promise spacetime stability, but fire, air, water, atoms, electrons etc, do not. One may attempt to impose ownership onto fire, air, water, etc, but this would be a futile gesture.

4.5 FREE AGENTS

Let T be an agent that we *a priori* consider to be no agent's property. The default state of any agent may be considered to be its own property[12]. An agent A begins in a free state. We may express this by any of the following conventions:

$$\pi_{1\,\text{free}}(T) : T \xrightarrow{\ +\text{owner}=*|\pi_{\text{conventions}}\ } * \tag{4.17}$$

This convention means that T is initially owned equally by everyone. One could imagine labelling every free agent with a blank name, but this seems unnecessary. The more parsimonious alternative is that no one owns it:

$$\pi_{2\,\text{free}}(T) : T \xrightarrow{\ +\text{owner}=\emptyset|\pi_{\text{conventions}}\ } * \tag{4.18}$$

Finally, our preferred convention is that it owns itself initially:

$$\pi_{3\,\text{free}}(T) : T \xrightarrow{\ +\text{owner}=T|\pi_{\text{conventions}}\ } * \tag{4.19}$$

In all cases, it is assumed that:

$$T \xrightarrow{\ -\text{conventions}\ } \text{Society.} \tag{4.20}$$

and we recall that the naming of agents is a short hand for the promise

$$A_i \xrightarrow{\emptyset} * \equiv A \xrightarrow{\ +\text{name}=i\ } *. \tag{4.21}$$

so one could imagine labelling a thing T owned by some name like A simply as T_A or $T(A)$ etc.

4.6 VOLUNTARILY OWNED AGENTS

Now let's complete the promises needed to make ownership a mutual act of cooperation. An agent can make a certain promise if it has sufficient internal attributes or capabilities to be able to represent and communicate the promise in a form or encoding perceivable by others. If an agent can make such a promise, it can also promise to be owned by another agent.

An agent 'thing' T can promise its ownership by X, by autonomously promising it, assuming only that it has the capability to represent this label information on itself:

$$\pi_{1\,owner} : T \xrightarrow{owner=X} *. \tag{4.22}$$

Alternatively, if it has insufficient resources to remember 'owner=X', it could simply point to its owner with a typed pointer in such a way that all agents were in scope of the promise (e.g. a dog runs to its owner):

$$\pi_{2\,owner} : T \xrightarrow[scope*]{owned\ by\ you} X, \tag{4.23}$$

This is the so-called matroid construction[Bur14]. An agent that can neither point nor label its owner cannot itself promise to be owned or be the property of another agent.

The complete promise bundle has the form:

Definition 17 (Ownership by mutual cooperation).

$$\pi_1 : T \xrightarrow{owner=A} *, \tag{4.24}$$

$$\pi_2 : A \xrightarrow{-owner=A} T, \tag{4.25}$$

$$\pi_3 : A \xrightarrow{+I\ own\ T\ \mid\ owner=A} * \tag{4.26}$$

$$\pi_4 : A \xrightarrow{-owner=A} * \tag{4.27}$$

i.e. in π_1 the thing is labelled with its owner, in π_2 the owner accepts that promise, in π_3 the owner promises its ownership conditionally on the existence of the label, and in π_4 it assures others that the label exists and is accepted. This bundle of promises may be designated as the bundle $\Pi_{owned}(A, T)$, and effectively appoints the pair $\{A, T\}$ as a single superagent.

The ownership bundle in the box above is the (+) promise of ownership expressed cooperatively between owner A and thing T, but we still need the acceptance of the audience $*$, so there is still a matter of the following promises to A and T

$$* \xrightarrow{-I\ own\ T} A \tag{4.28}$$

$$* \xrightarrow{-owner=A} A, T. \tag{4.29}$$

We call this voluntary ownership, because the promise is formally given by the autonomous agent. This does not address the issue of what incentive or coercion might have been applied to encourage the promise, only the documentation of sufficient information to establish the resulting intention. Indeed, in the next section we could understand such a promise as the marking of an agent, e.g. cattle branding.

Just as the audience acceptance of cooperative association defines a superagent role as 'dressed agent', so the centre of ownership defines a cluster of property.

Definition 18 (Estate). *The resulting superagent cluster formed from ownership promises by various T, pointing to A could be called the estate of A. It is a* role by association *(4.3.1 in [BB14a]).*

4.7 CLAIMING OWNERSHIP
(ACQUISITION AND APPROPRIATION)

We understand claiming as taking ownership of something by imposition. An agent, which merely claims another without labelling it, cannot expect other agents to know about its claim. If the thing T has no capability to be labelled or branded on its body (i.e. is unable to make an owner promise), it therefore has essentially four options:

1. An agent A may try to impose a change of label onto a thing T, assuming this is possible. See section 4.8

2. Agents can use the services of a registrar (e.g. as in marriage). See section 4.9. A registrar is an agent that offers a 'ledger service' which keeps records of ownership information. This kind of record keeping is essential to accounting of goods, property, and money. We shall have much use for ledgers in connection with banks and money of account.

3. An agent A can attempt to hold on to its claim T so that others cannot make competing claims, until such a time as other agents promise to accept its promise claim of ownership.

4. An agent A can try to envelop the agent T with some kind of container, which can be labelled as its own. To do this, the agent A forms a superagent cluster of the container agent and the thing T, which can represent property instead, e.g. oil in a barrel, beer in a bottle.

It might not generally be sufficient hold a thing within an agent's interior. A subagent may only have the status 'borrowed' or 'held'. e.g a coin found on the pavement, a stray cat.

If the contained agent rejects its 'ownability' (see section 4.10), the claim can be refuted. Claiming of property might thus be considered a form of attack (an imposition), particularly in the case where an agent rejects ownability. However, if, an agent is ownable and there is no claim of previous owner, then it is *a priori* free, and may be claimed by imposition. Once held, an agent is on the interior of its owner's superagent boundary, where other exterior agents may not be in scope of its promises, rendering them unable to assess the ownership (and perhaps the existence) of the thing T.

Example 9. *If society permits an agent to be owned, by registering a claim with an accepted registrar, then it may not so easy for the owned agent to be able to deny its own ownability, because a third party can misrepresent the claimed agent, violating its autonomy. This effect might well be responsible for the confusion in some cultures about the assumption of ownership of spouses as property through marriage.*

4.8　Claiming (imposing) ownership directly

The following representation is a straightforward starting point[Bur14, Bur15, Bur16b]. Both agents are initially free, in the eyes of convention:

$$\pi_{\text{free}}(A_i) : A_i \xrightarrow{+\text{owner}=A_i \,|\, \pi_{\text{conventions}}} * \tag{4.30}$$

$$\pi_{\text{free}}(T) : T \xrightarrow{+\text{owner}=T \,|\, \pi_{\text{conventions}}} * \tag{4.31}$$

Agents begin as no other agent's property, and thus (from the rule of agent autonomy) it would be impossible to acquire another agent as property unless it explicitly promised to accept a change of ownership.

Definition 19 (Imposed claim). *An imposition by A_1, in the presence of a promise of ownability, results in a change in T's label.*

$$\left(\Pi_{ownable}(T) \;,\; A_1 \xrightarrow{+\textbf{def}(owner=A_1)} \!\blacksquare\, T \right) \; leads\; to \; T \xrightarrow{+owner=A_1} *. \tag{4.32}$$

Here $\textbf{def}()$ refers to the definition of a promise, not the body of the promise itself: this is how we write the proposal of an intention from one agent to another[BB14a]. So, in its own state, T now promises to be owned by A_1. To fully claim the agent, A_1 should also accept the ownership with the first of the promises (4.33) below, and further promise its conditional ownership given acceptance by T:

$$A_1 \xrightarrow{-\text{owner}=A} T \tag{4.33}$$

$$A_1 \xrightarrow{+\text{I own} T \,|\, \text{owner}=A} * \tag{4.34}$$

analogous to the voluntary case. The second promise (4.34) is a general declaration of ownership to all agents, making its claim public. This is optional, of course, but perhaps common practice. A_1 would also hope that other agents promise to accept the new state too, by making a promise analogous to (4.33). This is possible since the promise of ownership is made to all agents in scope, who are also potential owners, but it only has a functional value if there is some reason for the other agents to acknowledge the state.

This initial claim by imposition assumes the existence of a default promise bundle Π_{ownable}, i.e. that agents offer no resistance to being claimed. It implies that agents may acquire property by assertion, where there are no prior ownership claims. The question of a's free will, in this matter, is then pragmatically tantamount to the ability of a to remove or resist (not keep) the default promise π_{ownable}, which documents its implicit acceptance of this violation of its autonomy.

Example 10. *Can I have a napkin? No, I took these napkins for myself, get your own napkins!*

4.9 CLAIMING OWNERSHIP VIA REGISTRATION

In the case where an agent cannot (or will not) represent the promise of ownership π_{claimed} (4.42), a claiming agent might attempt to register the claim with a third party or *property registrar*.

Definition 20 (Property registrar). *A Trusted Third Party, maintaining a ledger L of 'deeds' and 'claims' on the ownership of distinguishable agents.*

Agents, which are not distinguishable cannot be registered (or rather a registrar would be deceiving its trusted base by accepting a claim) as they would lead to likely contention. The registrars promise is a promise of the second kind. Its trustworthiness is thus contentious, as it does not promise first hand knowledge; it makes an intermediate registration R or $L \in R$, corruptible, or potentially incorrect second hand knowledge.

Definition 21 (Ownership claimed by registration). *Let L be the ledger of a registrar agent, and A be the claiming agent.*

$$\pi_1 \quad : \quad A \xrightarrow{+I\ own\ T} L \qquad\qquad (4.35)$$

$$\pi_2 \quad : \quad L \xrightarrow{-I\ own\ T|proof(T),\pi_{unique}(T)} A \qquad\qquad (4.36)$$

This only makes sense in an environment in which T is a distinguishable entity:

$$\pi_{unique}(T) : T \xrightarrow{+locally\ unique\ name} * \qquad\qquad (4.37)$$

The second of the promises (4.36) is made conditionally here, but this depends on the registrar. Agents claims might be accepted on thin evidence, or only after a complex chain of custody (as in a blockchain record). These are completed by

$$\pi_3 \quad : \quad A \xrightarrow{+proof(T)} L \qquad\qquad (4.38)$$

$$\pi_4 \quad : \quad L \xrightarrow{-proof(T)} A \qquad\qquad (4.39)$$

What happens next is more contentious. The ledger now promises this trusted claim and makes it available to all interested parties:

$$L \xrightarrow{+A\ owns\ T} * \qquad\qquad (4.40)$$

Agents, who accept this promise (on trust), assume that the criteria of proof are sufficient, and that the trusted party has done its homework.

Example 11. *In the case of a marriage, where there is often a chauvinistic asymmetry in many cultures, it is easy to see that the function of a marriage registrar simple replaces 'T is owned by A' with 'T is married to A'. The structure is otherwise identical, and so it is not altogether surprising that the asymmetry leads to the blunt interpretation of marriage as a form of ownership of one agent by the other. Although we are not personally happy with this particular inference, it remains a virtue of Promise Theory that was can uncover such semantic patterns and their possible outcomes.*

Why would a registrar R maintain this ledger service? One reason could be that this is simply its nature (such as with a machine). At a human level, most likely there would be some exchange or remuneration for the service. If private, there might be a registration fee; if public, it might be a public service funded by government. Finally, we may note that the function of the registrar is essentially the same as that of a bank ledger.

4.10 OWNABILITY: WHAT CAN BE PROPERTY?

An agent unable to make or represent a promise cannot promise to be owned by another agent, or change its default to refuse the imposition of ownership by another agent. We assume that the following minimum promises must be keepable by an item that can indicate its ownership to other agents.

Assumption 3 (Ownability minimum promises). *Agents which cannot express promises to*

1. *Name their owner explicitly.*

2. *Uniquely identify themselves to a third party*

cannot be property.

This follows from (**Ax1**) and (**Ax2**). The converse is also true. Any agent, which promises both its owner and its identity is owned (see definitions 16 and 21). There are two ways an agent might be claimed, but only the first of these can be considered owned.

- If the agent T is capable of promising its owner (e.g. by name tag, branding, etc) then the matter is clear, then it can autonomously promise ownership (**Ax1**).

- If the agents are unique and distinctive, or can be *held*, by a container, which in turn can be identified, then ownership of these agents can be promised to a ledger of ownership, calibrated by some Trusted Third Party to whom all agents promise to accept as a source of binding 'truth'. The third party may promise to inform of such claims. In this case we may call it claimed or imposed.

Social convention admits the latter possibility, even for agents that are too primitive to be labelled, such as oil or water or other natural resources.

Example 12. *Generic, indistinguishable oil and water cannot really be owned. They can be held in barrels which can be owned, thus claimed. As long as they remain inside the barrels, they may be considered part of the property, but if they leak out, they become indistinguishable from anyone else's oil or water.*

The capability of an agent to reject the idea of its ownership might plausibly be identified with the existence of its capacity for freewill, though we shall not pursue such issues here. We denote this by assuming that the default state of agents is to promise to accept a redefinition of this ownership state from any imposing agent (see 3.5.1 of [BB14a]), with a bundle of two promises:

Definition 22 (Ownable agent). *An agent is ownable if it can autonomously promise the assumed characteristics of i) ownership, and ii) acceptance of ownership by an agent other than self.*

$$\Pi_{(ownable\ by\ X)}(T) : \{\pi_{claimed}(T), \pi_{claimable}(T)\} \tag{4.41}$$

where

$$\pi_{claimed}(T) \quad : \quad T \xrightarrow{+owner=X\,|\,(\mathbf{def}(owner=X)\,\wedge\,\pi_{conventions})} * \tag{4.42}$$

$$\pi_{claimable}(T) \quad : \quad T \xrightarrow{-\mathbf{def}(owner=X)} *, \tag{4.43}$$

The second of these promises leaves open the possibility of a change of ownership at any time, and the first promise automatically accepts such an imposition. A sufficiently smart autonomous agent could delete these promises; thus, rejection of such promises could be viewed as a simple criterion for distinguishing between A_i and T, i.e. agents that can and cannot be property.

Lemma 5 (Agents which cannot be property). *An agent T, which is unable to make the promises*

$$T \xrightarrow{+owner=X\,|\,\mathbf{def}(owner=X)} * \tag{4.44}$$

$$T \xrightarrow{-\mathbf{def}(owner=X)} * \tag{4.45}$$

may not be considered property.

This follows from (**Ax1**), (**Ax2**), and (**Ax3**). If T cannot delete its promises, it is not fully autonomous, and we can only treat it as an independent quasi-autonomous agent. Imposing ownership on an autonomous agent violates the assumption of autonomy, so an agent could reject the imposition or change its mind. The would-be owner X could further try to override the refusal to accept its imposition by force, but the intent to reject the ownership remains an autonomous act[13]. For example, the agent may not be able to represent the identity tag of the owner, or may forcefully reject the placement of such a tag.

Example 13. *Is innovation ownable (e.g. patents)? Can someone claim an idea? Can someone claim the uniqueness of a portrait, and deny others the right to paint the same picture? Is someone's time investment ownable (IPR) as Intellectual Property Rights? From the foregoing discussion, it seems clear that the documents of patents can be owned, and their contents by the same token. However, an idea cannot be labelled without its documentation container. Its representation in other forms cannot be claimed unless it can be shown that the idea was sourced directly from the owned documentation. In this case, similarity is insufficient to show causation or intent to steal.*

4.11 SHARED OWNERSHIP AND CLASSES OF OWNER

This definition may be generalized to the joint ownership by a number of agents, by replacing the owner A with a superagent cluster of agents $A_{super} = \{A_1, A_2, \ldots\}$ who all agree to share ownership of T.

$$A_{super} \quad = \quad \{A_i\}, \quad i = 1, 2, \ldots \tag{4.46}$$

$$A_{super} \xrightarrow{\pm terms} A_{super} \tag{4.47}$$

$$T \xrightarrow{+owner=A_{super}} * \tag{4.48}$$

This mutual agreement between the agents allows them collectively to be viewed as a single entity, or black box. All of the the promises clusters in the foregoing sections may be extended to refer to a list of owners, instead of single one. Owners may not have equal shares of ownership, and may thus fall into classes of priority or influence.

Example 14. *One could discuss Ltd stock/share companies. Companies often promise different classes of ownership to different shareholders: so-called preferred shares promise special terms not promised to regular shareholders.*

4.12 TRANSFER OF OWNERSHIP

Setting aside the possible need for remuneration for now, we can describe the transfer of promises (deeds) during a transfer of ownership. The promises might be formal documents, implicit conventions, or practices documented in law.

Transfer of ownership from A_1 can be accomplished by emission of the item T, which is simply a release of ownership back to a free state (with no new owner) or a transfer to a new agent A_2. Note, it is a separate issue is whether A_1 expects some remuneration for this transaction, or whether it is offered as a gift. We shall not address this here.

1. Emission (release) of an agent T from the body of agent A_1 to a free state (agent owned by itself T), involves the following changes. A_1 directs ownership to T replacing the promises in equations (4.33) and (4.34) with the following, including an imposition to change the owner of a back to itself.

$$A_1 \xrightarrow{+\mathbf{def}(owner=T)} \blacksquare \quad T \quad \text{(impose change)} \tag{4.49}$$

$$T \xrightarrow{+owner=T \mid \mathbf{def}(owner=T)} * \quad \text{(implemented change)} \tag{4.50}$$

$$A_1 \xrightarrow{-owner=T} T \quad \text{(accept change)} \tag{4.51}$$

$$A_1 \xrightarrow{+owner(T)=T} * \quad \text{(optional)} \tag{4.52}$$

$$A_1 \xrightarrow{+T} * \quad \text{(emission)} \tag{4.53}$$

This is analogous to spontaneous emission in physics.

2. Emission (directed) of an agent T from the body of agent A_1 to target agent A_2 involves the following changes. A_1 deletes the promises in equations (4.33) and (4.34), and replaces them with the following, including an imposition to change the owner of T to A_2.

$$A_1 \xrightarrow{\;+\mathbf{def}(\text{owner}=A_2)\;}\blacksquare \quad T \quad \text{(impose change)} \tag{4.54}$$

$$T \xrightarrow{\;+\text{owner}=A_2\; \mid\; \mathbf{def}(\text{owner}=A_2)\;} * \quad \text{(implemented change)} \tag{4.55}$$

$$A_1, A_2 \xrightarrow{\;-\text{owner}=A_2\;} T \quad \text{(accept change)} \tag{4.56}$$

$$A_2 \xrightarrow{\;+\text{owner}(T)=A_2\;} * \quad \text{(optional)} \tag{4.57}$$

$$A_1 \xrightarrow{\;+T\;} A_2 \quad \text{(emission)} \tag{4.58}$$

This is loosely analogous to stimulated emission in physics.

3. Absorption from A_1:

$$A_2 \xrightarrow{\;+\mathbf{def}(\text{owner}=A_2)\;}\blacksquare \quad T \quad \text{(impose change)} \tag{4.59}$$

$$T \xrightarrow{\;+\text{owner}=A_2\; \mid\; \mathbf{def}(\text{owner}=A_2)\;} * \quad \text{(implemented change)} \tag{4.60}$$

$$A_2 \xrightarrow{\;-\text{owner}=A_2\;} T \quad \text{(accept change)} \tag{4.61}$$

$$A_2 \xrightarrow{\;+\text{owner}(T)=A_2\;} * \quad \text{(advertise change)} \tag{4.62}$$

$$A_2 \xrightarrow{\;-T\;} * \quad \text{(acceptance)} \tag{4.63}$$

These follow principally from (**Ax2**). If the initial owner imposes ownership, this is a transfer of ownership. If the final owner imposes ownership, this is *appropriation* of the agent.

4.13   DISAGREEMENTS ABOUT OWNERSHIP

Agents A_i may not respect a promise of ownership by other A_j, and may try to impose their own appropriation of T, as if T were still free. This then leads to contention over the ownership, which may or may not be supported by promises documenting the state. The promises described above may be made in the public scope to avoid contention by others. This is where social convention, and trust, may play a role in resolving the contention. The ability to remember payments, and register with an impartial or Trusted Third Party are a common solution. The ultimate belief in ownership is an assessment of whether certain promises are assessed as kept by all the agents in scope. This is assessed principally by the owner as an interested party, but might be disputed by any other agent.

The promise of ownership stems ultimately from an initial imposition by fiat or decree. Regardless of whether things found are assigned a legal (socially acceptable) owner on a first-come-first-served, basis backed by the social conventions, or some other measure of entitlement, ownership is only a belief that may be disputed. The ability for simple agents to promise allegiance to a uniquely labelled master is entirely beyond the capabilites of elementary agents, e.g. water, to keep. Thus stable ownership of any agent, without sufficient internal memory, is only a matter of imposed assertion, bolstered from dispute perhaps by the deterrent of a promise to defend the right of ownership with some force.

The observations documented above concerning property will be central to defining money, since money is involved in justifying changes of ownership. Why would be pay someone for something if we could merely take it? If ownership means nothing, then money is useless. In the case of services, where nothing tangible is transferred, we also try to argue this through 'intellectual property', or by outcome. Ownership is first imposed on a free agent T by claiming it, then later promised through transfers in a ledger (e.g. blockchain, logbook, name tag).

4.14 WEALTH AND ITS CREATION

Most accounts of money consider the idea of wealth, indeed greed, as a natural motivation for economic behaviour. It may be true that humans and other organisms have some kind of instinct for acquisition of property, but we do not strictly need to concern ourselves with the philosophy of motivations for holding or owning things in order to analyze the functional aspects of money.

Wealth is usually understood an accumulation of owned assets. This may be applied to an agent of any scale. This is crucial because the ownership of something by an individual is quite different to the ownership of something by a collective or network of agents[14]. Nevertheless, our definitions apply in all cases, in a scale-free manner. The concept of wealth and its meaning are indeed subtle (see [Bei05] for an excellent review). One can easily identify a number of different functional aspects of wealth semantics, (see section 5.14).

Definition 23 (Enabling wealth). *The ability to promise certain assets allowing the owner to pass certain obstacles that hinder its behaviour in some way, e.g. entry fee, down-payment for a house, access to a car, or access to unlimited flights.*

Definition 24 (Invested wealth). *The ownership of certain assets that could be promised in trade, in another context, where they give access to new possible acquisitions, or the return on the investment would be larger than if purchased directly.*

Definition 25 (Symbolic wealth). *The ownership of assets that symbolize social status, and perhaps grant access to exclusive clubs, services, or collaborations (i.e. like Darwinian sexual selection).*

Wealth is clearly not only about assets, and it is not about exclusivity; it is more about context within a network of promises. The usefulness of owning a certain thing depends on the environmental conditions in which an agent and its things find themselves. Wealth must therefore be evolutionary and emergent. Making gold watches is not creating wealth unless there is an opportunity for gold watches to play a role in the network of interactions, no matter how much effort goes into their creation. Even the terminology 'demand for watches' is simplistic from the perspective of economics as a so-called complex adaptive system[Art09, Bei05]. Here the association of wealth creation with work, by Karl Marx and Adam Smith was a distortion of seeing value through the lens of a factory production era.

Is wealth creation the goal of economics? Symbolic wealth is not a primary goal for a society as a whole, but enabling wealth almost certainly is. Wealth cannot be created without making a promise, which is assessed to be of positive value in a context of agents that can use the promise. This may include having access to other prerequisites.

Example 15. *A stockholder of tyres is not wealthy or valuable to a car maker who has no provider of wheels. Networks underpin wealth.*

Example 16. *Is value designed or does it emerge? This depends on the scale at which we examine systems and subsystems within them[Bur15]. A fashion designer might intentionally design a bag that is immensely popular and sells well, but she does so only in the context of a network of agents who are unintentionally attracted to this intentional act (they are semantically compatible). We cannot deny the existence or importance of intentional behaviour within the network, but intent is constrained by an evolving environment of unintended behaviours.*

Example 17. *External contextual circumstances may amplify or reduce the value of something. This makes it clear that money and value are unrelated. As value declines, no one pays us compensation. The value is simply lost. Later, if we resell something, how could we sell it again if money represented its value? If its value has not increased, why would someone else pay more for it? Did it acquire additional value? In fact its* *value may go up and down. Old whiskey gets more expensive. Retrograde fashion trends make old things new again. Value may be lost and found, but the money remains the same. In traditional economics, one tries to capture this through demand curves in a very simplistic way.*

If scale plays a role in the impact of intentionality, then we must also ask: at what scale does the averaging of promise semantics over a cluster of agents wipe out the significance of the intentionality, leading to the expression of a random variable? Scaling is not only about making an influence bigger or smaller, it is about the effect of overlapping and interfering promises, combining local and global constraints on outcomes. In the natural sciences, perhaps with the exception of chemistry and biology, we go to great length to avoid the intrusion of semantics and intent: by stripping away anything contextual or specific, we are able to compress general explanatory power into a few raw principles (a form of data compression). This is a memory strategy: the fungibility of explanation depends on how little context specificity it can remember. We shall see the same strategy in money, where universality is enabled by the indistinguishability of money in different contexts.

4.15 TIME LIMIT ON OWNERSHIP, AND PAYING FOR SOMETHING AGAIN

As soon as something exists, the clock of withering entropy starts ticking relative to thing and its environment. There is semantic relativity, but absolute degradation of the ability for a thing to keep the promises it originally made, without maintenance, over time.

Conversely, it is a widely held fiction that paying for something grants us full access to it for all time. Consumable resources and perishable resources, of course, have to be replenished. Houses and major items of ownership often need maintenance too. If we trap non-ownable items, such as a stock of fish and they escape from containment, we might have to pay twice to acquire them. Paying for goods and services may only grant us a kind of licence to claim ownership for a limited window (see example 18 in section 4).

Example 18. *Ownership may have a time limit. In China, for example, a person may only own property for 70 years (100 years in Hong Kong). After this time, the property reverts to the government.*

Example 19. *Ownership of copyright expires after 50-100 years depending on the jurisdiction.*

CHAPTER 5

MONEY

Equipped with this minimal formalization of ownership, expressing a social convention about the partitioning of both animate and inanimate things, we are ready to address the modelling of money and its attempt to conditionally reconnect them. We begin by describing money with a minimum of reference to speculations of value or utility (see sections 7.3 onward). Indeed, money and value present as totally different kinds of entity: money is a quantitative *promise*, measured in some units, while value is an *assessment* made by other agents, which need not be limited to monetary matters, or use monetary units.

We shall pursue this approach in the effort to understand money and value in a simple and consistent way, referring to:

- Amounts or measures of money (promises),

- Proxies for money (agents),

- Value or utility assessments (assessments).

Any theory of assessments is a theory of relativity. Money, on the other hand, promises an absolute artificial system of units for measurement, like standard weights and measures, which is (at least in principal) universally convertible into any equivalent currency, and *a priori* neutral in its affinity for particular goods and services[15]. While prices and valuations may change from day to day, the numbers on our coins, notes, and bank accounts are fixed in their promised amounts. This is the nature of a measuring stick[16].

5.1 ASSUMPTIONS ABOUT MONEY

All money begins by the issuance of promises. The promise to offer remuneration, the promise accept monetary compensation for some good, service, or future outcome. There are also the promises to accept agents of monetary transfer (coins, notes, cheques, bank orders, giros, data transactions, and so forth), all measured and approved as a standardized form of 'legal tender'. Money works as long as all agents within an economic area keep to more or less the same set of promises. Money, like property, then is a social convention.

In the literature, it has seen said that money serves three roles (see for example [KW13]):

- As a medium of exchange, i.e. an intermediary in the exchange of goods, services, and investments currently available.

- As a store of 'value': it is assumed to be non-perishable, with future purchase power, so it can be stockpiled in a mattress of a bank account for future use.

- As a unit of account: we use it to set prices in standard currency units, allowing us to compare items of different types in a single universal scale, however simplistic that may be.

Each of these functions makes assumptions that could be contested to some degree. The attribution of all these functions to a single concept seems to result in a loss of resolution, and some confusion about the dynamical behaviour of money. We shall recover the meaning of all three by clarifying the composition of these functions in terms of agents and promises.

A fourth function, which reveals the beginning of the rift between money and value, has been emphasized by Hart[Har00]: the acceptance of money embodies the idea of *trust*, and its strict accounting is a memory counter for past favours and transactions. Monetary balances summarize numerous social interactions between parties, as known from iterative game theory[Axe97, Axe84, BB06b, Bur13b].

Assumption 4 (Money accounts for trust). *We pay people back for goods, services, and loans, not because the world would fall apart if we didn't (as in the case of energy), but because the act of reciprocation builds trust. The accounting of money acts as a distributed ledger of exchanges, whose memory capacity is implementation dependent.*

If we can only remember a current balance of payments, then money has a very low resolution aggregate form of memory. If individual transactions can be maintained and recorded in a log, then semantics can be retained, even as money itself mixes as an aggregate sum. Trust measures the expectation of predictability, i.e. of keeping promises.

If one can rely on agents to be predictable, this is valuable. We could try to assess how valuable these favours and trades are in real terms, but a simpler time-saving approach is to attach prices and indubitable accounting measures that can keep track of the balance of payments. A simplistic agent might assess trust in another agent as maximal when its balance of payments was precisely zero, back and forth.

The three functions above, on the other hand, assume a semantic stability that is fictitious. An agent with money always has to ask: how do I know that this money will be accepted in the future? In dire political times money has indeed been perishable, currencies have disappeared completely without alternative compensation, banked savings have evaporated during financial crises. The changing acceptance of money leads directly to varying exchange rates and varying prices, thus these changing levels (which cannot be ignored) reflect a memory of the processes by which we arrived at the present economic state. So, in fact, if money stored 'value', as claimed, it would be considered a noisy and unreliable channel, in the sense of information theory, and one that frequently corrupted its storage.

Example 20. *Since money carries only an amount, which can be added or subtracted to existing ledgers or piles, it has no memory functionality. The history of interactions is rather more usually stored at the distributed end-points of the monetary network. This does not rule out the possibility of creating money with its own autonomous memory, of course (modern cryptocurrencies have this property), but this must remain implementation dependent. The common forms of money we experience in daily transaction, at the time of writing, are memoryless, i.e. they are Markov processes.*

If value were the true meaning of money, then, the fixed numbers on coins and notes would have to be considered lies from one day to the next. The path dependence of value has an important consequence for the assumption of value as a conserved quantity, an issue we expand on below. Money needs the appearance of conservation to make sense of accounting, but its usefulness varies like an umbrella in the tropics. In short, conservation of value would require path independence, or no memory of previous states (see [Mir89] for a philosophical discussion, or any advanced physics text referring to the Cauchy theorem for the mathematical reasoning).

In summary of all such considerations, we must begin by separating the notion of *monetary measure* from *representable value*. Untangling these concepts is not too difficult, in the language of promises, and we shall work through the issues systematically in the remainder of this section. Let us begin with a fundamental assumption:

Assumption 5 (The intended function of money). *The fundamental purpose of money is to communicate (to any party) an agreed measure of exchange, which has a common and persistent meaning to all parties, and which uses a socially accepted system of units.*

These properties describe a network transport system.

What agents do with this ability to document their exchanges need not be defined here, but of course we usually want to buy or sell something. Notice how this definition overlaps with the four points above: it is about exchanges, its purpose is accounting and measuring, and its integrity of record implies a memory of what goes on. The key divergence in this starting point lies in removing all reference to the notion of value. Our base assumption underlines, on the other hand, the promise of money to act as an *interloper* (see figure 5.1), or network transport layer, in the exchange of goods and services. This implies a fifth issue, which has yet to be mentions: namely the role of time. Money allows the advancement or deferral of exchange in a variety of ways that manipulate time. As long as it is available, it removes obstacles that might prevent agents from proceeding with their activities. This suggests that money acts as a kind of lubricant, which prevents society from stalling, fitting with Keynes ideas (see the discussion by Krugman[Kru08]).

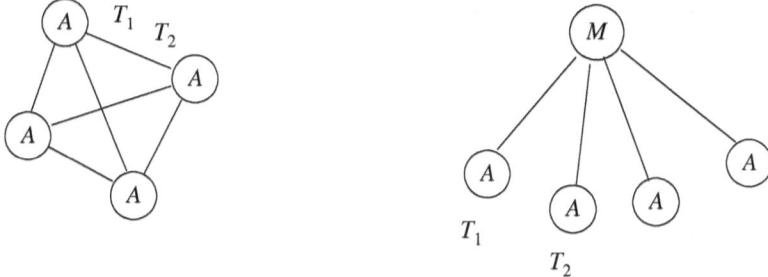

Figure 5.1: Money acts as an interloper (forming a calibrating matroid for exchange), replacing a peer to peer (p2p) network of direct exchanges based on individual trust (costing of order N^2 in the number of agents) with fidelity to a centralized source (hub) of measure calibration that involves only a single trust relationship per agent (costing trust of order 1 in the number of agents).[Bur14].

5.2 MONEY AS A NETWORK TRANSPORT SYSTEM

Why can't we simply trade goods and services directly for other goods and services (called barter)? Of course, we can, but this has major limitations. One reason, already mentioned, is time and space: it is a severe handicap insist on the immediate availability of goods and services in exchanges. There is also some controversy about whether there was ever a time in which this happened as the dominant form of trade; anthropological evidence suggests that it was a minority mechanism, but the definition of 'money' in these accounts is also quite liberal and varied[Smi04, Fer08, Har11, Suz17, Gra11].

5.3   SEPARATION OF PAYLOAD FROM TRANSPORT

To complete a physical trade, parties have to meet, discuss, and agree on the trade and then hand over the goods and services at one or more meetings. This might mean needlessly transporting bulky goods back and forth (e.g. along the silk road) at possibly inopportune moments, for every need. Money allows us to separate purchase from delivery, or defer payment, just as messengers and telegrams remove the need for parties to meet in person at precise moments of space and time. To decouple the physical transportation from the semantics of a trade we need some kind of invariant accounting. Consider the following example.

Example 21. *Suppose I have 20 goats and you have 11 cows. We might choose to exchange these, at a certain rate. By consulting the market forces (tea leaves, dung, star patterns, etc) we might divine that 15 goats are equivalent to 9.5 cows. We now have a problem. In this world of goods alone, we have now brought into being three currencies of exchange:*

- *Goats.*

- *Cows.*

- *Half-cows.*

Now, we have a question, which is not as trivial as it may seem. Do we consider cows and goats to be money in this example? We shall try to answer this question below.

In this a 'literal' exchange of goods, the traders need to have 'exact change' to be able to make the payment and the change needs to be available immediately[17]. We might choose to keep track of whole cows, but now we need a ledger to write down how much overpayment or underpayment has been made.

Whether or not we choose to physically divide cows into two halves, in order to pay exactly, in the foregoing example, there is nevertheless the *concept* of half a cow lurking in the transfer. It is impossible to avoid this, as humans are apt to place arbitrary and fractional assessments on things, unfettered by biological notions of completeness. Because neither cows, goats, nor the need for cows and goats are conserved quantities, universal or immutable, there is no reason to limit exchanges to these particular types[18]; rather, it makes sense to account for the exchange by postulating the existence of a neutral exchange parameter (which is analogous to the role of 'energy' in physics). In this example, we see the need to distinguish amounts from the carriers of the amounts, i.e. money from proxy. The limitations of a certain proxy technology could well limit the kinds of transactions that can be made. What is striking about these properties is their similarity to the promises made by network transport mechanisms.

We could define a quality of being like money. In [Key30], Keynes attempted for the first time to give a clear picture of money (see figure 5.4). Although quite dated now, the essence of his picture is still valid. There are two choices: to take an approach based on tradition[19], or on function. For the former, we define

---

Definition 26 (Moneyness or moneylike behaviour). *The degree to which a proxy/representation for money has the properties of commonly understood money in coins and notes.*

This is not without its problems: the original coins were made of gold and silver, and had a value, which led to problem like coin clipping (or goat clipping). So this scale has to define a kind of idealized view. Also, coins came before modern information technologies, where automatic ledgers (e.g. blockchain) could be attached to coinage. We might imagine fixing the scale of moneyness at 1 for coins, then values greater than 1 would add features, and values less than one would omit features.

Example 22. *The distinguishability of monetary proxies could easily affect their acceptability to certain users, while adding features like traceability. Users might refuse to accept money handled by terrorists, or 'tainted' by history, even though this very property would eliminate the possibility of money laundering. Users might even be stigmatized by the misfortune of holding money that was once held by an undesirable agent. This very fact could hinder the acceptance of certain payments, and prevent economic activity, as in a recession. Thus the universal fungibility of money might be lost by adding more information (like a BitCoin ledger) to money.*

Example 23. *China has been a pioneer in cashless electronic payment. In June 2017, Chinese Alibaba announced the creation of cashless cities on the Chinese mainland [Glo17, Yue17] in Fuzhou and Tianjin, based on visual QR codes and barcodes rendered on mobile phones, to transfer the information. In this case, one could argue that the money proxy is pure photons, through a non-trivial alphabetic data encoding.*

Proxy	Net	Own	Payload	Value	Mem'	Medium	Signed
coin	p2p	bank	fixed	metal	$O(1)$	metal	No
notes	p2p	bank	fixed	zero	$O(1)$	paper/plastic	Issuer
Bank MoA	hub	yes	variable	zero	$O(1)$	ledger entry	Bank
Check	p2p	yes	variable	zero	$O(1)$	plastic	Bank
Bank card	hub	yes	variable	zero	$O(1)$	plastic	Yes
BitCoin	p2p	yes	fixed	zero	$O(N)$	blockchain	Software
Airline miles	hub	yes	variable	zero	$O(1)$	points	Airline
Coffee card	hub	no	coffee cup	zero	$O(1)$	card	Franchise

Table 5.1: Some well known monetary proxies compared. The memory indicates how many previous owners money proxies can distinguish (1 implies only the current owner, as a Markov process). We have discounted the possibility that coins and notes may become collector's items, with speculative value, before or after their acceptance as money.

5.4   Network neutrality: avoiding preferential acceptance

The exchange amounts conveyed by money are *invariants* with respect to location, time, and all circumstances pertaining to an exchange. All changes and adaptations to circumstance, in quantitative numerics, happen through prices and payments (at the edges of the economic network, via the message rather than the monetary messenger). Prices form a set of promises that signal desired remuneration, reflecting changing circumstances of the agents in their specific context. As an interloper, money avoids contextual reinterpretation and circumstantial preferences. This is analogous to the concept of 'Network Neutrality' discussed in connection with Internet service provision[Wik].

If money promised any attributes to distinguish one transaction from another then any handling agent could choose to discriminate based on upon that information. Amount is already one possible attribute that allows preferential treatment, e.g. banks could prioritize large amounts or small amounts. However, if money remembered more information about where it had been, and to whom it belonged, etc, any party could act as a 'firewall' for preferential treatment.

Let us list a number of properties we consider to be pertinent to idealized money. These may be more or less represented by different proxies.

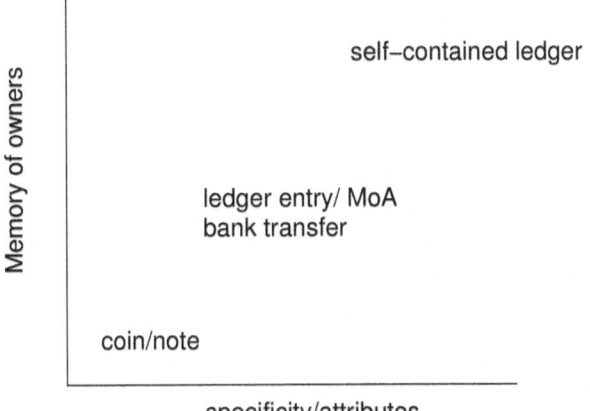

Figure 5.2: A rough placement of money proxies comparing their ability to represent detailed information versus the amount of memory or ability to recall specific attributes.

Definition 27 (Ideal monetary promises (moneyness)). *Necessary properties of money:*

1. *Promises a quantitative measure M.*

2. *Possessing no attributes or labels except its measure M (no intrinsic value and indistinguishable).*

3. *Agnostic to specific goods and services (fungibility)[20].*

4. *The units of quantity have fixed semantics assumed only for the purpose of buying and selling in all its forms.*

5. *Finite accuracy[21].*

We shall call this collection *classical moneyness* (or moneylike behaviour), to allow for the notion of future kinds of money based on different criteria. The neutrality of money has been essential to promoting its exchange usefulness in buying and selling, as an impartial measure of 'value in the moment', without making any promise about purchase power. Notice that ideal money's promises are scalar promises, in the language of[Bur15]: payments are vector promises, but ideal money itself points to no particular agent. The earmarking of funds for a particular purpose is a form of semantic adaptation of a kind of monetary proxy that can point to an intended recipient.

Example 24. *Earmarking of funds in a budget reduces their moneyness, according to these criteria. This makes sense, as it is a form of pre-spending, with the very specific* *intent to reduce the fungibility of the amount.*

The finite accuracy of monetary amounts leads to considerable subtlety, and makes the use of differential calculus contentious. Money is constantly subject to rounding errors, because prices and taxes are not constrained to the same denominations as money (often being represented as percentages, etc). This will ultimately prevent us from defining money mathematically as a metric space (see section 7.3 and onwards).

In order to count impartially, as an interloper, money should have no labels that can recall a relationship to a particular kind of good, service, or transaction type. All qualities, except numerical attribution, must be eliminated to make money impartial to transactions (like energy). We return to show for value this in section 5.15. Thus, in a similar way to energy, the story of money will be one (as for energy) of endless relationships to convert from one form into another, via a generic and universal bookkeeping quantity. In physics, matter, quanta, particles, and fields become the countable 'proxies' for energy. in finance, coins, notes, cheques, etc become proxies for money. Any agent which acts as a vehicle by which monetary value can be represented and transmitted.

Let us propose a couple of general design policies for money, which we frame as conjecture, since they are far from validated in any sense.

Conjecture 1 (Indistinguishability of proxies). *In order to avoid the impact of unintended information transfer during monetary exchange (payment), representations of money should have no labels other than the measures they represent. All qualities other than the semantics of the monetary units are superfluous to the intent of money.*

Today we already see, from credit card statements to transaction logs, that money of account is generally visible to the Trusted Third Party that mediates money. This has forensic uses, for example in law enforcement, if the privacy of the individuals is secured.

One sees quickly that money practically invents itself as a tool to avoid the awkwardness of accounting for different *types*—and yet, the invention of double entry bookkeeping adds it back on separate ledgers, for private use. In the pre-information age, it became expedient to have only a single currency of exchange over a wide area. However, this can also lead to too great a simplification: the money in the kingdom of Gilgamesh cannot be distinguished from the money in the neighbouring kingdom of Solomon, which neither king would approve of. So some labels seem necessary. The story of monetary proxies is the story of different technologies for recalling enough functional distinctiveness to separate concerns, while balancing against the need for fungibility relative to the users of a particular currency.

Conjecture 2 (Fungibility of money). *As a network interloper, ideal money is agnostic* *towards what it buys. The transference of monetary data implies no moral assessment or opinion about the reason for the payment.*

- This already suggests that money has different properties on the inside of a currency region and between different regions. So we might talk about interior and exterior money. This might even be suggested from the scaling of promises noted in [Bur15], where interior and exterior promises are key to distinguishing the scaling of agency and purpose.

- Money can 'get stuck' because it is typed. If money can only buy certain goods and services, and there is insufficient money of a different kind to buy something else, then an economy may not have enough money of all kinds in a typed economy to function.

- The liquidity of money will be important, because of the role of ideal money as pure information. Like coins that you put into the meter to keep the power going.

- Historically, the reason why money became the preferred form of exchange (to barter) was precisely its interchangeability – being able to put off reimbursement or exchange to a later time, or to have the freedom to convert the return into a form the bartering partner cannot offer. This decoupling of time, exchange, and change of ownership is an important semantic characteristic of money.

- At certain times money proxies may lose their money status. For instance, notes from the 1800s become collectors' items, and are sold for high prices at Southerby's. At this point, the proxy has lost its moneyness, and has become a 'good'.

Example 25. *BitCoin[BdL13] has zero value as information: its cost is the amount of CPU needed to generate the encryption keys of for the BitCoin, or so called 'mining' of BitCoin, but its value as data is zero. This makes BitCoin a free currency, but not a fair one, since only users with significant resources can create the money. The monetary authority is the authorized software.*

Is there a difference between money and goods for barter? If there were something that everyone always needed more of (say energy), would this do as a means of exchange? In earlier times, this was certainly the case: working for one's supper was a common means of sharing. In a work economy the ubiquitous need for labour and the fungibility of humans as a labour force.

What is striking about these properties is their similarity to the promises made by network transport mechanisms.

5.5   TWO STATES OF MONEY

Money exists in two states, corresponding to being held (deposited) or being in transit (in payment). Its promises or semantics are somewhat different in these two states, as we shall see. For example, as a payment, money acts as an agent of exchange through its proxy. Once absorbed or deposited with a host, it plays the role of a pure promise, which can be aggregated into clusters, where it can promise insurance against unexpected future events, etc. The assessed value of holding versus paying is thus context dependent, and leads to much hedging in the semantics of payment.

5.6   THE AGENTS AND SEMANTIC SCALES OF MONEY

Money can change forms or roles, but it also exists within a framework of other roles: goods, services, buyers, sellers, banks, and so on. All of these players are represented as agents in Promise Theory. We shall mention a few of these here, and illustrate their relative scales in figure 5.3.

We assume that 'blank' promise theoretic agents have no *ab initio* properties, except those that are explicitly promised (this includes the assumption of a physical representation). Agents may be mere concepts, quite abstract until realized. We shall assume that any such realization is something that can be promised by such a representation. Thus the value of agents lies only in the promises they implicitly make. Promises are formally made by agents, even if the origin of intent is only proxy for human interpretation.

Agents may cluster into 'superagents', which act and appear as a single agent to some entity at a similar scale. So, agents at any scale may make promises[Bur15]. At a scale S, one may imagine a collection of agents, associated somehow, as being surrounded by a fictitious 'boundary'. Promises that are entirely on the interior of the agent boundary play a different role to promises that cross the boundary or are made by the collective boundary itself[Bur15]. This has immediate implications for the semantics of money, and leads to a distinction between endogenous and exogenous money.

Thus promises are a faithful accounting principle for utility. The Promise Theory used here was introduced as a way of modelling policy semantics from the perspective of independent collaborations[Bur05a], and was later developed into a general theory of cooperation[BB14a, BB13, BB14b]. The scaling of agency was discussed in [Bur15].

Definition 28 (Bank). *An agent that creates interior money, stores exterior deposits, acts as a tenant for account customers, and plays the role of an network interchange hub. The bank also promises to exchange interior money for exterior deposits or 'cash', accessed through different platforms (coinage, notes, cards, cheques, transfers, etc). These functions include lending and financing to customers.*

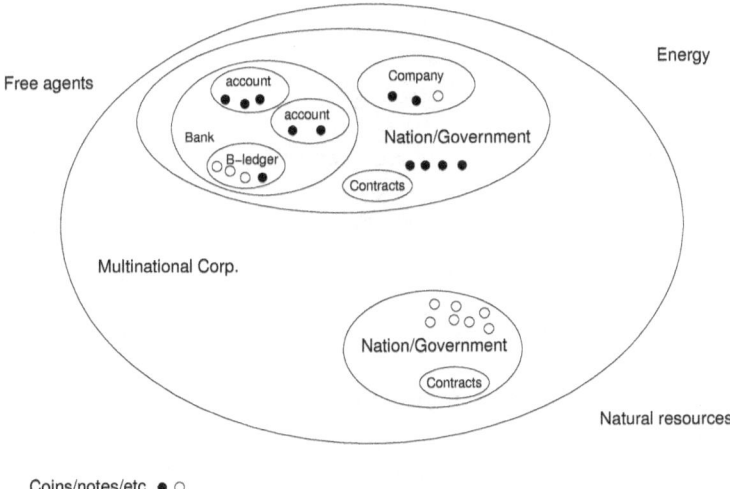

Coins/notes/etc ● ○

Figure 5.3: The scales of agency in an economy. The containment of agents has the partial order *currency/debt* < *account* < *bank/company* < *nation/government* < *multinational* < *unclaimed resources*. Note that coins and notes, as well as goods, may be located inside or outside of any of these levels; currency tender located outside of the economic agents is either lost of stashed.

It is a common impression in society that banks and government are somehow outside of the normal activities of buying and selling, but this is not the case. Banks, governments, and every agent in a society are always trying to be financially solvent and 'make a profit'. Money is a universal currency for interchange. The instrument for accounting money in the modern world is the bank account[22].

Definition 29 (Bank account). *An agent which promises to transfer and received payments, promises to record the balance of payments in and out. Each bank account behaves as a tenancy[Bur15] or rented space belonging to the bank, but whose contents are owned by the account holder, but situated within the bank, sharing common services.*

5.7 EXAMPLES OF MONEYLIKE PROXIES

Money has different representations and vehicles. We use the term proxy to indicate the vehicles for money. Keynes[Key30] sketched out figure 5.4 in the 1930s, before many modern monetary technologies were available.

Here we sketch out some of the promises made by different proxies:

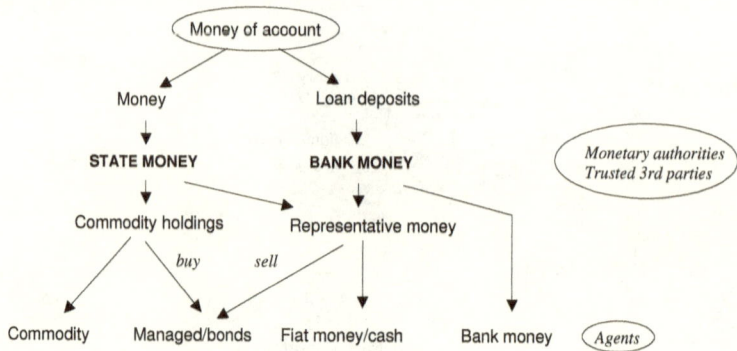

Figure 5.4: Keynes overview of money types. Notice how banks have been granted the power of government in the issuance of money.

Definition 30 (Cash). *Any representation of monetary measure that promises immediate availability for exchange (liquidity).*

Definition 31 (Coins). *Coins are a form of cash, made from authorized metal alloy, which promises a pre-authorized value, usually from the set*

$$M \in \{0.01, 0.05, 0.10, 0.2, 0.5, 1, 2\} \qquad (5.1)$$

Coins are issued by a mint and authorized by a central bank. Modern coins are atomic and indivisible[23]. Coins effectively promise ownership by the mint, but may be held by any agent. Coins have no transactional memory today, and leave no traceable record, of their holders over time; in the past, they could be clipped and assessed by weight.

Definition 32 (Bills and notes). *Bills are a form of cash, made from authorized paper, plastic, and foils, that promise a pre-authorized value, usually from the set*

$$M \in \{1, 5, 10, 20, 50, 100, 200, 500, 1000, \ldots\} \qquad (5.2)$$

Bills are issued by a mint and authorized by a central bank. Notes are atomic and indivisible. Notes are owned by the mint, but may be held by any agent. Notes have no transactional memory, and leave no traceable record, of their holders over time.

Bills, like the pound note, which 'promise to pay the bearer the sum of one pound' in equivalent measure of gold, from olden times, were promissory notes. Coins made no promise to be exchangeable at a bank, so they were formally a distinct form of cash. Today, the difference has no real meaning.

Definition 33 (Cheque/check). *A parameterized representation of money of account, which simulates cash, but without a pre-authorized value. It is validated by a retail bank rather than a national or central bank. Cheques are are issued by a private bank, pertaining to an individual account, and are authorized by the same bank. Cheques are indivisible, but may be transferrable. Cheques are issued in a blank state by a bank, their ownership is unclear. They may be held by any agent. Cheques promise validity only for a single transaction. They leave a record of their holder in the account ledger.*

Definition 34 (Money of account (bank money)). *A form of ledger or database over monetary amounts in and out. An account can be held by any legal entity (person, company, etc). Each account holder has its own ledger. What makes money of account especially flexible is that it coalesces into a total 'balance' and can be spent in arbitrary amounts. It can be created by loan, and it never leaves the bank. Money of account is owned by the account holder, and is held by the bank (the opposite of cash). It keeps a long term memory of transactions, and is therefore traceable.*

Definition 35 (Money transfer (transaction)). *Two ledger entries within a bank, or between banks to amend the money of account ledger entries of the owners.*

Definition 36 (Pledge). *A public promise to pay a monetary amount or non-monetary outcome in the future. This is a 'desired outcome' promise, possibly with conditions attached.*

Definition 37 (Loan or mortgage). *A form of debt, which may be paid over a long term, in multiple installments by contract agreement. A loan is owned, and may thus be transferred, or bought and sold.*

Definition 38 (Token). *A moneylike counter often made from paper or plastic, redeemable for something else. Tokens are specific to a locale.*

Definition 39 (Coupon). *A moneylike note usually made from paper, redeemable for something else in whole or in part. Coupons are specific to a locale, company, or collaboration.*

> **Definition 40** (Discount voucher). *A form of coupon issued by retailers that requires the spender to use some money along side.*

Example 26. *Many moneylike tokens exist today, as corporations have the financial muscle to acts as banks, without licences, or tax collection.*

- *Airline miles—moneylike tokens.*

- *Coffee cards—tokens, say 1/10th of a cup of any coffee.*

- *Buy one get one free.*

- *Discount on your next purchase of 'commodity'..*

- *Petrol stamps (Green shield, Coop)*

- *WeChat.*

- *Paypal.*

- *BitCoin.*

- *World of Warcraft money.*

The ability for money and its proxies to be aggregated and divided in deposited sums is what makes it fungible and universal. Theoretically, money could be construed as an atomic system, in which all amounts of money were aggregates of a basic minimum unit. This is true in payments, but not in expectations. Monetary amounts (prices, taxes, etc) are based on the real numbers, and do not respect this atomic structure of amount aggregation.

5.8   TIMESCALES IMPLICIT IN MONEY TRANSACTIONS

Although money is invariant in space, time, and all aspects of exchange, time plays a central role in the semantics of payment, and the promises made by certain monetary proxies. There are many sequences and timescales implicit in the exchange of money, and while these are almost never stated in economic theory, they will prove to be crucial to its understanding.

Time is closely woven into the semantics of money, because access to money affects the time at which goods and services can be bought. Money is also an exchange proxy for human interaction—a depersonalized intermediary that maintains a relationship bond. It is also a rentable asset, through lending. Thus holding money may be an advantage, and not holding may be a disadvantage. This advantage is usually quantified by an *interest rate*, which is a clock based on regular charges or payments for holding money owned by another agent (see section 7.15).

Definition 41 (Time deposits). *In a deposit account, this refers to money deposited into an account by the account holder, which cannot be withdrawn before a promised elapsed time*[24]. *One says the policy 'matures'.*

The ability to make exchanges *asynchronously* is a key function of network protocols. Payment with money enables such a protocol. How long before or after an exchange of goods should payment be due? How long do we have to pay a bill? After what time does an agent have to declare bankruptcy? Can it recover? Remarkably, in spite of our societal stigmas about bankruptcy, these issues are only accounting questions that have little to do with the need for real suffering. This is why companies regularly enter bankruptcy, only to emerge again at a later time after money is either injected, loaned under supervision, or simply written off with a legal mandate.

Example 27. *Bank account holders who have too much debt are often offered supervised accounts, to avoid bankruptcy. The bank ensures that money is used to cover obligations, taxes, rents, etc, then the remainder is offered in pocket money. In this way, customers are kept within a financial system and are given a chance to recover. This is like the bankruptcy or receivership terms for companies in many countries.*

There are many more cases in which money acts as a buffer to time criticality.

- Trades and transactions as ticks of the clocks belonging to parties with shared scope. These may or may not share a common clock.

- Time may be measured differently by sellers (+) and buyers (-) of stuff.

- Assessments of markets, prices, availability, etc, take a finite amount of time, and set an aggregate timescale.

- The schedule for interest computations by lenders sets a timescale for borrower jeopardy.

- The timescale over which changes occur to goods, services, e.g. spoiling, consumption, degradation, maintenance, replacement, etc.

- How long do we have to pay a bill, or to settle its debts? The timescale for cashflow accounting. When is an agent in the red, and when is an agent bankrupt? How long does the agent have to redress a negative balance before the time limit has been exceeded.

5.9   THE PROMISES INVOLVED IN MONEY

As we shall see, through the examples and separations of concerns, the essence of what characterizes money is to make a fairly simple promise, without complicated semantics. Money is a very elementary symbolic language (in the Chomsky sense[Cho59]) for communicating intent to purchase, in which the statements and utterances are the payments and transfers associated with buying, selling, or giving of gifts. This linguistic connection was first pointed out by the Turgot in 1769, but has since been overshadowed by other aspects of money[Tur69b]. As we'll see, the real language is price.

Any semantics, beyond this basic intent to transmit an amount, are normally promised 'out of band', as an independent contextual information channel, or imposed directly by the receiver. For instance, money may be payment in reference to an invoice or bill. It might be labelled as such, with accompanying information, such as a letter or note to the bank, an invoice number, or a payment ID. Since the channels are separate, they can easily be detached and estranged from one another. Thus monetary transactions can remember but can also forget.

Example 28. *We focus, as usual, on promises of the first kind in this work, but there is also a role for other kinds of promises when thinking about money[BB14a].*

1. *Promise of the second kind (to provide information):*

$$A[B] \xrightarrow{-M} C \tag{5.3}$$

A promises C that B will accept money proxy M, e.g. A tells C that a certain card M will be accepted in in a hotel B in another country.

2. *Promise of the third kind: (to provide a service)*

$$A \xrightarrow{-M} C[D] \tag{5.4}$$

A promises C that A will provide an amount of money μ to a friend (D) of C who will visit A in the near future. This may be helpful if D has problems when travelling with money proxies.

3. *Promise of the fourth kind: (by an intermediate agent that a service will be provided by someone else). This occurs in informal international money transfer, for instance.*

$$A[B] \xrightarrow{+\mu} C[D]. \tag{5.5}$$

A runs (or participates in) an informal money transfer scheme (this kind of thing exists in the West for transfer to countries without adequate banking system)

A (at place P) promises C (also at place P) that B (at place Q) will make a payment of μ to D (also at place Q). For the transaction A promises to accept the same sum plus a fee ϕ from C:

$$A \xrightarrow{-(\mu+\phi)} C. \tag{5.6}$$

5.10 MEASURE

Definition 42 (Monetary measure (amount)). *A promise made by money concerning the intended volume μ of transference, as measured in the units of the promised currency.*

The currency may have a type, e.g. USD which has to match the imposed payment request.

Definition 43 (Money (monetary promise)). *Money is the promise (by some agent M_i) to imbue a definite quantitative measure for 'right to purchase'.*

$$\pi_{money} : M_i \xrightarrow{+\mu} * \tag{5.7}$$
$$\tag{5.8}$$

Money is an abstract concept. Its status might be likened to a mathematical group or algebra. It has no a priori representation, but there may exist many such representations, with possibly incongruent properties.

Definition 44 (Authorized or socially accepted money). *A specific instance or representation of money, authorized by a monetary authority, or accepted for exchange. It must be able to promise:*

1. *Convertibility to goods or services.*

2. *An identifiable measure for exchange.*

3. *An authorized 'signature' on its proxies (watermark, serial number, special material alloy composition, digital signature etc) by the authorizing monetary authority.*

5.11 MONEY PROXIES

Let us define money proxies more carefully.

> **Definition 45** (Money proxy or representation). *A generic term for any agent or collection of agents that promise to represent and act as proxy for exchange value. Let $\mathcal{M}$ be a collection of agents representing money (money proxies), and $M_i \in \mathcal{M}$ by a money agent. Then, if $*$ is any (or all) agent(s),*
>
> $$\pi_{money} : M_i \xrightarrow{+\mu} * \tag{5.9}$$
>
> $$\pi_{auth\ proxy} : M_i \xrightarrow{+S_{auth}} MA \tag{5.10}$$
>
> *where μ is a fixed monetary value, and S_{auth} is an authorized signature that promises verifiability by a Monetary Authority MA. The latter promise may or may not be visible to all parties; it only needs to be known to monetary authorities.*

Consider the superagent structure of a money proxy that carries money of measure M (see figure 5.5). Using (**Ax2**) and (**Ax3**), we can construct the superagent collaboration. If the amount represented is μ, then we shall refer to the exchange measure as $\mu|\mathrm{rep}(\mu)$ to emphasize that this is a separate assisted promise[25] that depends on the monetary promise. The situation in the figure may be written in notation:

$$\pi_{money\ amount} : \text{Money} \xrightarrow{+\mu} * \tag{5.11}$$

$$\pi_{money\ owner} : \text{Money} \xrightarrow{+owner_\mu} * \tag{5.12}$$

$$\pi_{accept\ rep\ amount} : \text{Money} \xrightarrow{+\mathrm{rep}(\mu)} \text{Proxy} \tag{5.13}$$

$$\pi_{amount\ to\ represent} : \text{Proxy} \xrightarrow{-\mathrm{rep}(\mu)} \text{Money} \tag{5.14}$$

$$\pi_{proxy\ amount} : \text{Proxy} \xrightarrow{+\mu|\mathrm{rep}(\mu)} * \tag{5.15}$$

$$\pi_{proxy\ attributes} : \text{Proxy} \xrightarrow{+other\ attributes} * \tag{5.16}$$

$$\pi_{proxy\ owner} : \text{Proxy} \xrightarrow{+owner_{proxy}} * \tag{5.17}$$

$$\pi_{proxy\ authorization} : \text{Proxy} \xrightarrow{+owned\ by\ MA} * \tag{5.18}$$

Promises (5.14) and (5.15) are the formal binding between a proxy and its monetary value, imprinted at the time of manufacture of the proxy. Holders of the proxy have to trust the authorization of this binding. The proxy relationship is a service relationship: formally, the proxy provides a service to represent the money client publicly. The proxy and the money need to cooperate in forming an assisted promise that assures correct communication of the amount. Note that the owner of the money (a client) and the owner of the proxy (e.g. Royal mint, or central bank) are usually not the same. Readers should also not fall into the trap of confusing intent by proxy with the difficulty of attributing the 'intent of a coin'; such algebraic formalities are as important to the accounting of consistent promise logic as proper bookkeeping entries are to monetary matters. In this example, the proxy plays the role of a 'thing' agent T as well as money. This dual role is what leads to a conflict of interest.

Definition 46 (Monetary authority). *Any agent MA, which promises to authorize or validate a money proxy or currency token M_i, or redeem the measure of the money to its holder H in an alternative form. In some cases, the authority may also own the proxy, i.e. the monetary proxies may be the formal property of the monetary authority MA (as in the case of the Royal mint).*

$$MA \xrightarrow{-S_{auth}} M_i \qquad (5.19)$$

$$MA \xrightarrow{-owner} M_i \qquad (5.20)$$

$$MA \xrightarrow{+valid \mid S_{auth}} *. \qquad (5.21)$$

where $valid(S_{auth})$ is a computable function of the signature S_{auth} The first promise is the complement of that given in equation 5.10 to inspect any monetary agent, the latter represents the promise to disclose the result of validation to any agent.

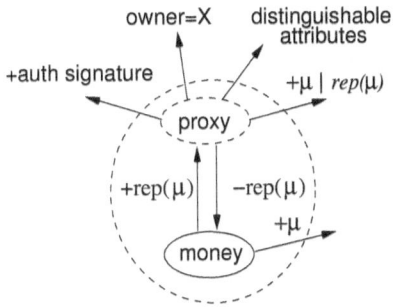

Figure 5.5: A money proxy cluster consisting of the money and the vessel. We see that the number of promises involved in proxying for money is non-trivial, even for the simplest coin. There is money represented by the coin, and the acceptance and authorization of the representation, as well as the coordination of the measure with the proxy's public appearance.

5.12 THE ACCEPTANCE OF MONEY AND ITS PROXIES

The promise of money is useless without a corresponding promise to accept it. There are two separate issues: the acceptance of the monetary amount (accepting the power to purchase in the future), and the acceptance of a proxy for money (the form in which it is transferred, e.g. cash, bank transfer, etc). One could also talk about accepting a container holding money, e.g. carrying someone's purse, but that does not imply accepting ownership of the money.

The acceptance of a proxy may not imply the acceptance of the money (e.g. here are some coins to pay for my coffee), just as wrapping a proxy in a holder does not imply acceptance (e.g. hold my wallet for me). Every agent may independently promise (or not) to accept money, recognizing its measure, and trusting its authority.

Definition 47 (Acceptance of money). *Let R be any agent accepting money from a sender S, and let M be any money proxy used to make the exchange. Agents R accept two things: i) that the proxy M represents actual money in the amount μ, and ii) the money itself:*

$$\pi_{accept\ money} : R \xrightarrow{-\mu} M \qquad (5.22)$$

$$\pi_{accept\ money} : R \xrightarrow{-\mu} S. \qquad (5.23)$$

Thus a recipient may accept the measure, and the validity of the proxy, without actually taking ownership of the physical proxy. This can be contrasted with the acceptance of a money proxy agent M by absorption:

Definition 48 (Acceptance of money proxy). *Let A be any agent, and M be any money proxy, and H by an agent currently holding the proxy. The agent A can promise to accept authorized proxies for money.*

$$\pi_{accept\ proxy} : A \xrightarrow{-M} H. \qquad (5.24)$$

The acceptance of the proxy does not imply the acceptance of the amount.

In any exchange, all these promises may be in play.

5.13 TRUST IN MONEY

The willingness to make these promises to accept money is rooted in an underlying trust[BB06b] of all agents involved. Trust, like utility, is conditional on the promise to accept (see section 2.1 and definition 2). This is implicit in the nature of promises made for the interpretation of human agents. From its network properties, it follows that the familiar network patterns in figure 5.1 also apply to monetary trust[BB06b]:

- *Trust in peers*

  When transactions are performed between peer agents, each agent needs to trust the other in an $O(N^2)$ network. This is an expensive, time consuming, and memory intensive process. Indeed, the Dunbar hierarchy has been shown to place limits on how many agents humans can form trust relationships with (for a review see [Bur13b, ZSHD04]). There is thus an 'economic' incentive to reduce the need for expensive peer trust.

- *Trusted Third Parties*

  By routing trust through an institutional agent, one relieves ordinary peers from the need to trust one another, replacing this with trust in the institution. Each agent who trusts the third party also effectively votes for it, and bolsters its reputation. This leads to a stable association. As we have observed in the financial crises of the 20th century and the 2008 crisis[For16, Kee17, Har00], trust is grudgingly robust to even rogue criminality, because of the great saving of individual responsibility, and perhaps a sense of there being no practical alternative.

  The Trusted Third Party that may be a government, a banks, an algorithm (as in BitCoin).

If there is a relationship between money and value, then it surely lies in its being a proxy for trust. The acceptance of money from someone, is a trust building communication. Trust is a form of memory, and money is a dominant form of communication that binds us into repeated interactions called trust relationships[26]. Banks and third parties grease the wheels of this communication, by simplifying the memory issue: we only have to trust the routers or third parties that isolate us from direct harm, but in a transparent way that does not form an obstacle to the eventual formation of peer trust too.

When trust is in short supply (to use an economic metaphor), one attempts to verify and make promises conditionally. The presence of additionally promised information alongside the monetary amount, e.g. a particular shape or design of coin, offers distinguishable criteria that could be used for filtering acceptance.

Lemma 6 (Non-money attributes allow preferential handling). *Any distinguishable promise made by a transferrable agent M (e.g. money proxy)*

$$M \xrightarrow[H,\ldots]{+b} * \tag{5.25}$$

allows the emitting and receiving or handling agents H to discriminate based on the promise, provided they are in scope of the promise, i.e. the promise is visible to them.

This observation is the basis of 'firewall' or access control technology in communications networking. Any discriminatory capability can be used both for defense or offense. Preferential handling of money can lead to economic obstacles and a kind of prejudicial handling, even ultimately economic warfare based on preferential acceptance and non-acceptance.

Example 29. *In some countries electronic payment systems require customers to have a local social security number, post code, or address in order to make a payment. This discriminates against tourists and foreigners who then have to pursue some kind of workaround, with its attendant impact on the tourist economy.*

5.14   THE VALUE OR UTILITY OF MEASURES OF MONEY

As we have argued, there is a distinction between money and value. However, we need to untangle the two from the confusion that arises when its holders perceive money as a power to acquire things perceived as valuable. We have emphasized that money has invariant measure, but that the value of things is a relativistic quantity, subject to contextual distortions. This allows us to ask: how valuable possessing money may be, to an agent, relative to its circumstances. This will give us a simple way to answer the question of intrinsic value in the subsequent section.

A simple way to assess the value of having money is through the ability of that money to overcome economic obstacles.

Definition 49 (Economic obstacle (dependency)). *Something that stops an agent from keeping another promise, for want of a dependency.*

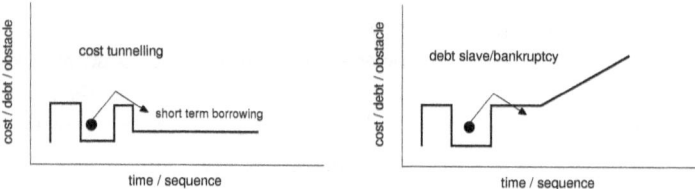

Figure 5.6: An economic obstacle can be represented easily in terms of a universal exchange measure. As the cliché goes, 'If everything can be turned into a measure of money, then there is no problem that can't be solved by throwing money at it'.

Holding the money is valuable provided it can be exchanged for surmounting the barrier. In this sense, we are proposing that the value is represented as a willingness to accept an amount of money by an agent A:

$$\Pr\left(A \xrightarrow{-\mu} *\right) \tag{5.26}$$

In fact, the latter is a less ambiguous term, and is preferred outside this section.

Assumption 6 (Partial linearity). *We assume that money has the semantic property, by convention, that there is a linear relationship between the perceived value of money* $v_A(\pi_{money})$ *and its measure* μ, *over a certain range of* μ. *In the context of some observer O, the value of an offer (+) may be written:*

$$v_O \left(M_i \xrightarrow{+\mu} * \right) \propto \mu \quad \mu_{min} < \mu < \mu_{max}. \tag{5.27}$$

Outside of these ranges money may lose its value, depending on context, but its measure μ *is immutable. This value is a piecewise function, its range dissected by the discreteness or indivisibility of goods and services.*

For instance, a sum of money that is too small may be useless to someone in need of a deposit for a loan, or in urgent need of expensive medicine. Conversely, possessing an already large measure of money, the value of more money to the recipient becomes reduced.

More generally, one might try to assume that the value of money is a general monotonic function $E(\cdot)$ of its measure:

$$v_O(\pi_{money}) = E(\mu), \quad E(\mu)_{min} < E(\mu) < E(\mu)_{max}. \tag{5.28}$$

In other words, the value of an amount of money might depend non-linearly on the amount, because of hidden semantics of the environment in which it acts. Only money above or below a certain threshold might entitle agents to access certain outcomes. What this indicates is that a system of money based on value would have unstable semantics, due to network effects. Certain properties of money might make sense peer to peer, but when placed in a network of interactions become unsustainable (see also from section 7.3 about payment).

Lemma 7 (Money is not a store of value). *Savings and reservoirs of money are invariant stores of monetary measure, but not invariant stores of value.*

The potential value of owning or using money is proportional but still independent of its monetary amount. We propose to abolish the use of the term 'value' in connection with money, because it leads to frequent misunderstanding[27].

Example 30. *Why then do we confuse money with value, when we don't confuse metres with length? The reason, of course, is that money is only a promise to update ledger entries at the bank, whereas value is subjective like trust (and unlike length). Just as users of the Internet have difficulty understanding the different between the network packets and the information they seek, so we are apt to muddle money with goods and services at the endpoints of the channel, even more so when we assess them as valuable.*

5.15   INTERFERENCE WITH THE MONETARY FUNCTION

The goal of a money proxy is to transmit a specific measure of money between agents. If the proxy has intrinsic value, this interferes with the measure of money transmitted by the proxy, because the proxy could be exchanged for money too (e.g. the gold coins could be melted down and sold for electronics at a higher price). Thus money proxies that are themselves valuable interfere with the intended function of the money, by creating a side-channel with alternative semantics. This may be perceived as a promise conflict. We can see this as follows.

Consider the exchange in (5.18), and figure 5.5. Let some present or future holding agent receiving or carrying the proxy be denoted by H, then we denote H's assessment of the value of the exterior promises made by the money proxy by these two statements:

$$V_H(\pi_{\text{proxy amount}}) \quad = \quad \mu_1 \tag{5.29}$$

$$V_H(\pi_{\text{proxy attributes}}) \quad = \quad \mu_2 \tag{5.30}$$

where μ_1 and μ_2 are both measures, in the units of money μ.

The value associated with the transfer of this proxy unit is now not single valued. If the receiver in a transaction is able to disregard one of the values, it might pick either μ_1 (exclusive) or μ_2 (let's say the largest of the values $\max(\mu_1, \mu_2)$), since it can only use one of the functions at a time (either by spending the money requiring the proxy intact, or by auctioning off the proxy and losing its monetary value). Note that μ_2 might actually be negative, if proxies were made of some toxic substance, like contaminated cash. However, if it does not think in such practical terms, it might imagine the value to be $\mu_1 + \mu_2$. In all cases, the value of the proxy μ_2 is open to speculation from buyers, whereas the value of the money it communicates μ_1 is fixed.

Lemma 8 (Optimal monetary communication). *The maximum certainty in money transferred by a single atomic money proxy occurs when $\mu_1 \gg \mu_2$, or $\mu_2/\mu_1 \to 0$, so that the limiting valuation of both $\max(\mu_1, \mu_2)$ and $\mu_1 + \mu_2 \to \mu_1$.*

Money should therefore have no intrinsic value ($\mu_2 \to 0$) in order to maximize the certainty of what is communicated by money, leaving only a single dominant channel of communication. This result is intrinsic and a direct result of the propagation by mutually promised binding. If money were merely a one way obligation or imposition, this would not be possible. A convenient way of expressing this is by defining a measure of efficiency.

Definition 50 (Informational efficiency of money). *The ratio of the perceived value of money to the total perceived absolute value of the money along with its proxy.*

$$\epsilon = \frac{|v_A(M \xrightarrow{+\mu} A)|}{|v_A(M \xrightarrow{+\mu} A)| + |v_A(M \xrightarrow{\emptyset} A)|}. \qquad (5.31)$$

When the values are automatically positive $0 \leq \epsilon \leq 1$. However, it is possible for the perceived value to become negative, e.g. when owning notes or coins is a liability, thus we write this in terms of the absolute value $|v|$. Any non-zero encumbrance associated with the intrinsic value of the agent reduces this efficiency for pure information transfer.

An ideal monetary vehicle or proxy has no intrinsic worth, which could interfere with the free transfer of information.

Lemma 9 (Non-zero value reduces the information efficiency of money). *If a monetary agent or proxy M_i has an intrinsic value $v_A(M \xrightarrow{\emptyset} *) \neq 0$, there is a corresponding influence on its acceptance by A.*

Even though we choose to put the faces of famous individuals onto currency notes, along with holograms and fancy patterns, we can transfer the value a single number with far greater efficiency. The high information content of notes is for authentication, not for monetary purpose.

We may ask: what if the promises $\pi_{\text{exchange value}}$ and $\pi_{\text{attributes}}$ were, in fact, completely indistinguishable? This is, in fact, impossible, because a proxy must be a physical agent, which must have exterior attributes (good or bad), so we would have to make its value fixed by altering the amount or composition of the proxy in real time, as the buyer's valuation changed (by weighing out gold power, for instance). This cannot be achieved with a fixed or invariant proxy agent, only with a composite bundle of agents, measured out one by one.

The latter point suggests another possibility for interference: that the amount of money transferred can, itself, interfere with the amount intended in virtue of its size. This follows from equation (5.28) above. Even if there is no intrinsic value in the proxy, there might be a value to holding onto a reservoir of deposited money, rather than using it for an intended transfer, e.g. as a future insurance policy, or get a better price at a later date. This is related to the time semantics of payment, and the two states of money in section 5.5. The reservoir or bulk accumulation of money has a value, which the individual proxies do not have, and thus there is a new form of contention between promised intent.

Example 31. *In hard times, agents may refuse to part with their money now, believing that prices may be cheaper in the future, but only if they know they might have enough to invest a minimum amount.*

So monetary value is also intermingled with *time* in a fundamental way. Such speculation about the future exchange potential of money may lead to unexpected obstacles if one of the possible channels for using the value in the future became blocked, e.g. a recipient could no longer use the money, but could sell gold, because of privileged access to markets.

5.16   INTERFERENCE FROM TRANSACTION COSTS

The bare amount of money is its promised amount μ. This can be 'dressed' by a veil of transactional charges and additional encumbrances: taxes (VAT), levies, interest payments, etc. All of these alter the local functional efficiency of money to purchase goods or operate as an investment. The role of these charges in the larger picture is impossible to describe without knowing the full network of intent[28].

Example 32. *Transactional charges may include bank charges for:*

- *Administration.*

- *Staff employment.*

- *Cost of communication and computation infrastructure.*

- *The cost of holding property, e.g. for storage of cash reserves.*

- *Insurances.*

- *A surplus savings plan.*

The addition of new side-channel semantics to money increases the potential for interference of intent (see figure 5.7).

5.17   THE CREATION AND DESTRUCTION OF MONEY

So far we have assumed that agents have money already, but we have not accounted for where this money might come from, or if there is enough of it. Like any other technology or commodity, it must be manufactured. It is true that money is given in return for labour (work), but this cannot be what creates it. Employers too have to get it from somewhere. They get it from customers, who get it from them, and so on. Labour might produce goods and services, but the money to buy them needs to come from somewhere else, in a closed network. So where does money come from? There is an empirical answer to this question, and a theoretical answer.

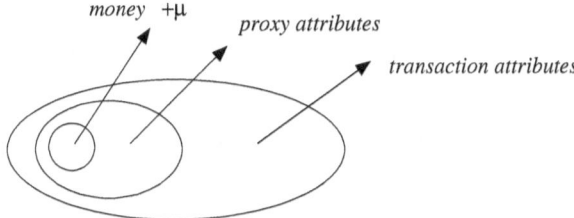

Figure 5.7: The nesting of promises in transactional wrapping is analogous to network protocol layers in computer engineering, and adds potential interference at each level, because there is no clean separation of semantics between the function of the layers, due to the possibility of allowing users to exploit the information channels in time and space for a change in potential monetary gain (gambling behaviour to attempt the influence of future outcomes).

In any economy money is created by 'fiat'. It is posited, imagined, minted, printed or written down, by some agent A. For any self-appointed monetary source A to create money, it is necessary and sufficient for it to promise the existence of some monetary proxy M:

$$A \xrightarrow{+M} * \qquad (5.32)$$

The movement of money in the network of customers (which may include the banks) assumes the immutability of money, or the integrity of the ledgers, as codified in the assumption of *homogeneous accounting*.

Assumption 7 (Homogenous accounting). *Trust and acceptance of money relies on the assumption that all agents handling money account for it correctly, without loss or markup. Only authorized agents may create new money.*

In the economy at the time of writing, certain agents are exempted from this rule; these are principally banks. At several places in this work, we refer to the 'conservation of money' as meaning that money does not appear or disappear by magic. What this really refers to is the assumption of homogeneous accounting. Money can indeed be created or destroyed by 'magic' (or at least by fiat), but agents are expected to refrain from doing this on the grounds of a notion of fair[29] distribution of wealth, which in turn maintains a sense of trust and social cohesion.

Although necessary and sufficient, the promise above is not useful. Money is only useful if it is accepted by at least one more agent. Theoretically, money can be created by any agent (e.g. a different currency for each transaction), but in practice this is not done, for two main reasons:

- The trick is to get your money accepted by others, based on the belief that others will be able to exchange the money for goods, with you or other agents in the community at another time and place, thus maintaining a fair distribution of things. Other agents will not usually trust money to fulfill its function without some imagined guarantee, which is provided by a legal regulation[30].

- It is (at least initially) parsimonious and efficient for all agents to use a single form for all transactions.

The licence to promise a system of money, for public use, is usually regulated by law, principally to build trust in it, on the assumption that potential users trust the law: a licence to issue money is promised conditionally by some appointed monetary authority, based on certain promises being kept by the recipient[MRT14]. Typically, a 'central bank' promises private banks a licence to create money subject to certain conditions (see figure 3.3). Attaching conditions to the creation of money has a number of useful side effects.

Example 33. *Authorized agents (banks, in the modern world) create money on their ledgers or balance sheets through lending. Thus the principal condition for creating money is that it will be paid back. Customers can move this money around, but cannot create anything new, no matter how hard they work or how much gold they dig up. The central bank may issue cash directly and may destroy cash. A central bank cannot go bankrupt[All11]. Everyone else in the network of agents promises, one way or the other, to preserve the integrity of money. This is only a promise. It cannot be guaranteed. The entire financial edifice is based only on these promises.*

Why don't we all make our own money? The two reasons given above, are compelling but not final. In fact many companies, who hold a sufficient buffer of assets to absorb temporary redistribution of those assets, do indeed create private tokens whose validity is limited to their company's scope. Since money is only a promise, any agent may promise its own money. The trick is to get others to accept it. Money is a network operation, so it is useless unless it connects you to someone else, and can be used to acquire something we want. The ability to act as a network service provider for the economy is thus a generally job for large licensed carriers (like banks). A sufficient number of agents probably need accept someone's kind of money, in order for it to perform its trusted function—because the support of many agents signals trust and stability and lowers the bar for accepting it. If we all made our own money, it would be analogous to everyone speaking their own language, albeit quite a simple language.

Any agent who can attract a sufficient following creates money by writing down the information of an amount on a ledger. It might issue authorized tokens that represent this fixed amount, for mobility, or it might redirect others to register for an account with them and simply move the amounts around on ledgers (as in banking). In any event, nothing other than documentation is created, except perhaps for the tokens, notes, coins (which we have already established in section 5.15 should have no intrinsic value).

Example 34. *In antiquity, tally sticks were used to make monetary transfers. A male and a female image could be matched, like yin and yang to confirm the authenticity of marks on the tally[Gra11]. These correspond to + and - promises in a promise exchange, and indicate how simple personal accounting can work as a form of money, credit, and debt tracking.*

5.18 MODUS OPERANDI OF MONEY CREATION TODAY

In present day society, it is mainly banks that are monetary authorities, licensed to create legally authorized money. This requires a banking licence with attendant regulation. State law, in most countries, regulates banks to ensure that they don't simply invent as much money as they feel like, and to award it to themselves; however, these regulations fluctuate. The laws in the United States, for instance, were relaxed in the 1980s to allow banks much leeway in doing precisely that. Of course, what private companies or individuals do in the privacy of their own homes, campuses, or inside computer systems, games, or other closed environments, is not the business of government, or law (at least for the time being), so in fact anyone can create their own money for internal distribution of goods and services already owned within that border. This is now well established in computer gaming and closed social environments.

The circumstances of money creation, in the capitalist society, are roughly as follows:

- Private banks create new money by loaning money.

- If a central bank wants to increase the money supply, it buys a financial 'asset' or instrument (e.g. a bond or security, or any multitude of other names) from the bank and pays for it. This works in the same way as the private bank: it is a pure ledger operation. When it wants to take money out of circulation, it sells the 'asset' and the money is repaid by the bank. All this means is that the authorized amount of ledger money the bank is authorized to hold of its own or others' money is now reduced. The asset is then, in principle, possible to sell to someone else, but their money would have to come from some other bank[Ind11]. Central banks thus make money by printing it, or by buying fictitious 'assets' or 'securities' from private banks, and they destroy money by recalling it from circulation, or by selling the fictitious assets back.

- Interest charged on the loans, which is a separate topic, acts as an incentive to repay quickly. Without this incentive, there is no fundamental reason to repay a

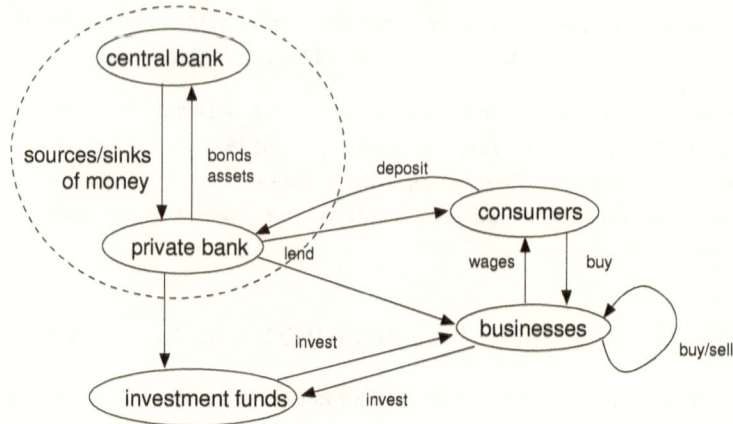

Figure 5.8: The flows of money from banks to consumers and businesses. All profits and new things have to be made or bought from existing money, moving around. The total amount of money cannot increase without borrowing.

loan, since no one would be actually inconvenienced if something imaginary were not taken seriously. However, there would be a loss of trust, due to a sense of unfairness. The status of loan repayment is more analogous to a test of character, and the assessment of credit-worthiness is considered to be a licence to borrow or acquire money.

- If banks can simply create money in this way, one might ask why they could not simply create an infinite amount of money. The main reason is that they have to obey their terms of licence with respect to a central bank. A central bank can give and take money from a private bank by buying assets from the bank. Circumventing this 'fair system' would violate trust and lead to a potential collapse of the currency. Another reason is that having too much money in circulation is assumed to lead to greed, needless buying, and 'inflation' of prices, thus eventually negating the usefulness of the money.

This last step is interesting, from a promise theoretic viewpoint, because it has the nature of an obligation to sell, rather than a promise. The asset is created by the private bank, but the central bank decides when it is going to get its money back. Without such a system, banks might never have to give money back, and this could lead to an oversupply of money.

In summary, one sees that no (new) money can exist that is not simply conjured into being by an authorized agent, exchanging a debt for a deposit, on its ledger. Our commonplace notion of the physicality of money, as something valuable like gold and silver, is simply wrong, in the modern world[MRT14]. The physical tokens of cash are only proxies that may be used as a means of small scale mobile exchange (outside of banks).

CHAPTER 6

BANK AUTHORIZATION AND MONEY REGULATION

Promise Theory tells us that, in order to calibrate the semantics of banks and money—make money homogeneous and universal for all its users—banks can most cost effectively calibrate their intentions by subordinating to a central agency[31]. A centralized authoritative role avoids the slower and more costly $O(N^2)$ process of mutual equilibration, which could be harmful to users, just as long as satellite agents of the central hub accept subordination. This is the role of a central regulator, often a central bank. When banks interact with foreign banks, however, there is no global oversight, so it would still be possible for entire countries to defraud one another. Indeed, some 'tax haven' countries have lax licences that do not properly regulate the monies. For governments, moving money through authorized channels allows the possibility to collect tax more easily.

Independent regulation is needed to verify that promises are being kept. The simplest approach would be the use of an independent third party regulator. This could be an institution or even a software standard. The alternative is to use peer oversight, based on consensus to calibrate transactions. In Europe, the Basel auditing standard for banks promises a standard of calibration[bas14] with tight controls. Although detailed, these are not hard to accomplish in the age of information technology. Indeed, information technology becomes the essential enabler for scaling modern financing. However, peer oversight can also become a cartel for illicit behaviours. In the computer age, the fairness of dealings could be taken out of the hands of people altogether by computers. Software applications acting as intermediate agents can be monitored and automated, and regulated cheaply. With the addition of a payment agent, like a debit card company, or PayPal, WeChat, Alipay, Apple pay, etc, there would be three intermediaries[32].

The role of banks in calibrating trust offers a simple explanation for the rise of centralized accounting and money of account (see figure 6.1). Banks worked as Trusted Third Parties when everyone used the same bank. With competing banks, the bank-to-bank phase of the above transfer via intermediary banks simply reverts to the case of peer to peer transfer (the first case), which is completely unverified. Again, why should we trust this? The cost of peer to peer consistency is of $O(N^2)$, while the cost of centralized calibration is only $O(N)$. Banks are expected to promise regulators that they will not

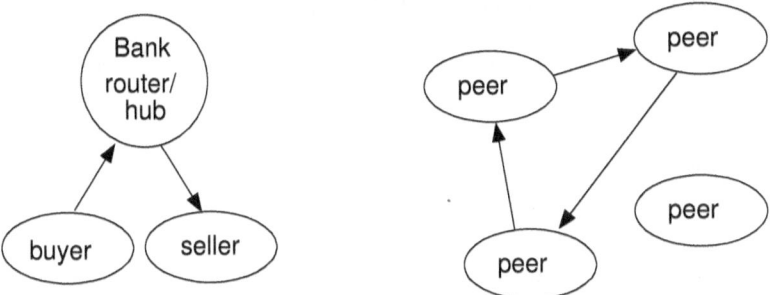

Figure 6.1: Money allows a balance of payments to exceed simple pairings of agents, and behave like a network. As usual, there are two ways this coordination can be managed: as a single central hub which acts as a calibration agent (banks), or peer to peer with emerging stability (as in the case of cash, and international currencies). The advantage of a bank hub is that money never has to leave the bank.

promise more holdings than a multiple of liquid cash reserves, since customers should be able to withdraw their liquid deposits on demand (though these rules were violated during the financial crises[Kee11]). This is only required in the case that there is 'run on the bank', i.e. users suddenly demand their account balances in cash. The rules allow them to lend up to a certain multiple of their total deposits in loans. In practice, with the deprecation of cash, this means little. More importantly, banks need to attract a trust community, else their money is impotent. It is their balance sheet in total that enables them to function in a useful way. Thus a positive rate of interest on deposits acts as an incentive. Thus the regulations often include promises of the form:

$$\text{Bank} \xrightarrow{+\text{reserves} > R\% \times \sum_i \text{Deposits}_i} \text{Regulatory authority} \qquad (6.1)$$

$$\text{Bank} \xrightarrow{+\text{transaction reporting}} \text{Regulatory authority} \qquad (6.2)$$

The Basel regulations include many more promises that banks are expected to keep, to maintain their licences[bas14]. It remains unclear the extent to which this can truly be assessed however. By exchanging money and transporting cash, banks could lend each other deposits to overcome any perceived limit on the level of deposits. In practice, controlling the precise amount of money in circulation seems at best impractical.

Lemma 10 (Strong oversight). *Unless banks have strong binding to regulatory observer, they can create money without limit, ultimately undermining trust in the banks' ability to keep promises, and thus their money.*

6.1   BANK ACCOUNTS AND LEDGER AGENTS

To account for banks, we introduce two subagents of the bank: a typical customer account A_C, and a bank ledger L (which represents the bank's own internal bank account).

Definition 51 (Bank account). *Bank accounts are tenancies, sold for rent or that promise ledger services and terms and conditions.*

Definition 52 (Ledger). *An agent L that promises to recall amounts of money $\mu(L)$ available to its owner, and a history of the transactions emitted and absorbed over since $t - T_{ledger\ horizon}$ and the current time t.*

The amount of information recalled by the ledger depends on the capability of the ledger agent. This is a promise made by the ledger. If the only promise is to recall the current value, then the agent has the Markov property in transactions

$$L \xrightarrow{+\mu(L)} * \tag{6.3}$$

If L can additionally recall a journal of past transactions, we can write:

$$L \xrightarrow{+\mu(L),T_1,T_2,...,T_n} * \tag{6.4}$$

The loss of transactional information may be considered a kind of entropy, as the system forgets its past. A ledger can, in principle, retain this forever.

Lemma 11 (Account memory). *Ledgers can remember who paid money into it, and to whom outgoing money was paid to. However, this memory is not retained when it gets transformed into a form of money proxy that is incapable of being owned. Cash, coins, and notes are untraceable.*

Lemma 12 (Loss of type memory). *Ledgers do not record money proxy types. All money is converted into money of account in a single currency, thus all origins are lost (entropy).*

Some bank accounts promise to retain money as multiple currencies, meaning that the ledger has a limited memory of origin or exchange. However, the key promises of amount, and other associated semantics, are almost entirely dissociated for the common money proxies, but not for debt. Debt carries with it all kinds of history to trace risk, and impose obligation.

Example 35. *The bank accepts deposits of cash. From the bank's perspective, any money held is simply a commodity. Indeed, normal cash has no memory capability, and could have come from anywhere. Banks don't typically accept shoes, flour, whiskey, but may accept property in payment for defaulting on promises. There is no reason, in principle, why banks could not take deposits in money tokens, air miles, collectors stamps, or any other form of transactional proxy memory, and act as a more general ledger service. Indeed, modern social media platforms are starting to do this, especially in China (WeChat, Alibaba, etc).*

Example 36. *Virtual banks, e.g. Skandiabanken, Sbanken in Sweden and Norway, operate in the same way, but have no independent infrastructure for money. They embed themselves hierarchically—such that the whole bank is a single business account of the host bank, held as an account agent in another bank—and divide up that account into subagents, representing individual customer accounts (see figure 6.2).*

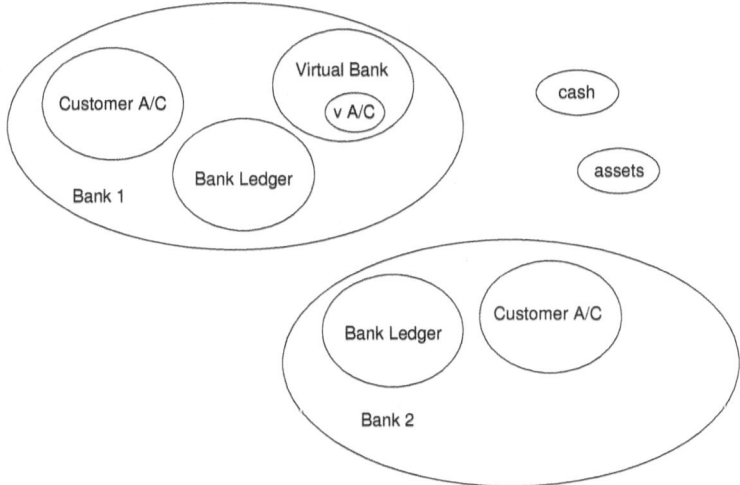

Figure 6.2: The agents and subagents in banks.

6.2   BANKS AND DEBTS (MONEY OF ACCOUNT)

In most cases today, money of account has the status of a service, provided by a monetary authority. Sometimes authority is centralized; in other cases, it may be de facto non-central, but all money plays the role of a network go-between that originates from some kind of hub.

Lemma 13 (Money of account is a service). *Money of account is a shared conditional ledger service provided by a bank, in which account holders can transfer monetary amounts to one another, assuming a positive balance.*

The proof is straightforward from definition 6 of a service. The promise of ledger entries on imposition of a payment is honoured conditionally by the bank, e.g. given that there is a minimum balance. There is a service provider (a bank or bank account), which promises a ledger service to users from the bank agent.

We assume that banks and their customers act as autonomous agents. The autonomy of these parties immediately implies that a bank cannot create money without a customer promising its generic desire for a loan, with amount μ_C and currency C. A loan is therefore a conditional promise offer, in response to that desire:

$$\text{Customer} \xrightarrow{+\text{loan-request}(\mu_C,C)} \text{Bank} \qquad (6.5)$$

$$\text{Bank} \xrightarrow{-\text{loan-request}(\mu_C,C)} \text{Customer} \qquad (6.6)$$

$$\text{Bank} \xrightarrow{+\text{contract}|\text{loan request}(\mu_C,C)} \text{Customer} \qquad (6.7)$$

The customer makes its readiness known, (**Ax3**), the bank engages by accepting this interest, and offers a contract in response.

Definition 53 (Loan contract promises). *The terms of the contract may be a quite long bundle of bilateral promise proposals, to be made real by signing, including:*

- *The amount of money offered* (μ_B, C)

- *The method of interest calculation.*

- *A rate of interest, or agreed source.*

- *Conditions of repayment, installment schedule, termination, etc*

- *The promise to impose a sum μ_B of money onto the customer's account (and agent called 'Customer account'), and the promise to accept it.*

Both agents sign the contract to agree to its terms[BB14a], and reify the promises:

$$\text{Customer} \xrightarrow{\pm sign_C} \text{Contract} \qquad (6.8)$$

$$\text{Bank} \xrightarrow{\pm sign_B} \text{Contract} \qquad (6.9)$$

$$\text{Customer} \xrightarrow{-contract(\mu_B, C)|sign_B, sign_C} \text{Bank.} \qquad (6.10)$$

The acceptance of this contract now effectively represents the creation of money, since it is a necessary and sufficient condition for the creation of a deposit. Banks create money, they do no simply reinvest other people's savings. Having agreed to the terms, the signature is usually taken to imply acceptance of the loan offer too.

Definition 54 (Bank Loan). *A service provided by banks, by which money created by a bank, for repayment at a later time.*

The loan is a contractual agreement (see section 8.4 of [BB14a]) between two parties, to honour the deposition of M currency units for immediate (liquid) access, subject to the terms and conditions agreed.

$$\text{Contract proposal} \xrightarrow{+terms} \text{Bank, Account holder} \qquad (6.11)$$

$$\text{Bank} \xrightarrow{+\mu|terms} \text{Account holder} \qquad (6.12)$$

$$\text{Bank} \xrightarrow{-terms} \text{Contract proposal} \qquad (6.13)$$

$$\text{Account holder} \xrightarrow{-terms} \text{Contract proposal} \qquad (6.14)$$

$$\text{Account holder} \xrightarrow{-\mu|terms} \text{Bank} \qquad (6.15)$$

$$\text{Account holder} \xrightarrow{+sign_C} \text{Bank} \qquad (6.16)$$

$$\text{Bank} \xrightarrow{+sign_B} \text{Bank, Customer} \qquad (6.17)$$

Once, agreed a bank imposes the loan transaction to make the deposit and debt registration, as promised in the contract:

$$\text{Ledger}(L) \quad \xrightarrow{+(A_C \to A_C + \mu_B)|\text{terms},\text{sign}_B,\text{sign}_C} \blacksquare \quad \text{Account}(A_C) \quad (6.18)$$

$$\text{Account}(A_C) \quad \xrightarrow{-\text{contract},-\text{sign}_B,-\text{sign}_C} \quad \text{Ledger}(L) \quad (6.19)$$

$$\text{Bank} \quad \xrightarrow{+(L \to L - \text{Debt}(\mu_B, C))} \blacksquare \quad \text{Ledger}(L) \quad (6.20)$$

where A_C is the balance of the customer account and L is the balance of the bank's ledger. Note that the promise of debt is a $+$ type promise, even though the money amount is formally of negative value, because it is holding the debt, not accepting it. The ledger and account, being owned are assumed to accept such impositions by default

$$\text{Customer account}(A_C) \quad \xrightarrow{-(A_C \to A_C + \mu_B)} \quad \text{Bank Ledger}(L) \quad (6.21)$$

$$\text{Bank Ledger}(L) \quad \xrightarrow{-(L \to L - \text{Debt}(\mu_B, C))} \quad \text{Bank} \quad (6.22)$$

These formal statements may be regarded as schematics of the machinery of banking. We assume that L is owned by the bank, and that A_C is owned by the bank, but rented by the customer (as a form of tenancy[Bur15]).

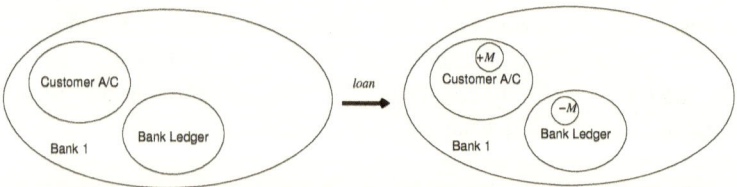

Figure 6.3: The creation of money by banks, is simply a book-keeping shuffle: we write $0 = +\mu + (-\mu)$ and separate the halves, like virtual matter-antimatter creation in physics.

The accounting, represented by these transactions, only makes sense if money is conserved (see figure 6.3). It also shows that the conservation of money is a *purely voluntary act*, which is quite difficult to regulate. It depends entirely on the goodwill of agents involved. Unlike the elementary agents of atoms and subatomic particles that do not have variable degrees of freedom to behave irregularly, people and banks have far too many to even keep track of. Conservation of money assumes that the bank, creating the money, voluntarily accepts the debt as both liability to itself and to the customer, and that receipt of monetary deposits will cancel this debt[33]. Note that it could fail to keep either of these promises, and violate the conservation of money. These are not self-evident necessities. A bank, especially one that is unregulated, can easily create and destroy money, with only its reputation at stake if the information were to spread.

6.3 MONETARY DEBT IS A NON-MONEYLIKE PROXY

The concept of debt is important to understand how banks treat money of account. Money of account is ordinary money, indistinguishable from any other money once paid into a bank account. However, something interesting happens when money of account is created by loaning money from the bank that cannot be understood in terms of quantitative balances alone. To understand this, we need to clarify the semantics of debt.

Definition 55 (Debt). *An assessment, by a promisee, of the measure by which a specific promise has not been kept, measured in the units of the intended outcome.*

Definition 56 (In debt). *A state, attributed to a promiser, in the lifecycle of one of its promises, during which the promise is incompletely kept, i.e. the outcome is not fully discharged.*

A monetary debt is thus a monetary measure in an outstanding balance of payments, which is assessed by the potential recipient (payee). Because debt has specific semantics, and is attached to a specific distinguishable promise, it is not simply 'any old money'.

Lemma 14 (Monetary debt is a non-moneylike proxy for money). *Debt promises an amount $\mu_{balance}$, a payer agent A, and a label that distinguishes it from other debt $\pi_{payment}$.*

Pure money is a singular promise only of an amount μ. This is insufficient information to track the semantics of debt. Monetary debt is associated with the incomplete remuneration of a particular promised amount. This will be important because the act of creation of money leads to a specific intentional event, with labels to that intent that must be preserved to complete the trust accounting (each separate promise was paid in full, without prejudice). It can only be carried in a proxy with minimum requirements, and labelled with specific intent.

Definition 57 (Repayment (of debt)). *A monetary promise $+\mu$ with the intent to reduce debt amount by targeting its specific ledger entry $Debt \rightarrow Debt - \mu$, i.e. as opposed to payment of interest, fees or charges associated with the debt.*

The amount of money referred to in a debt thus has to be distinguishable from other monetary amounts, and has a registered owner, maintained on the ledger of the promisee or payee. It fulfills the criteria for a proxy, and it is earmarked with a specific intent. So it is more than simply money. By (**Ax2**), such a targeted payment also has be be accepted as such by the debt holder. Failure to accept repayment may have critical consequences for trust, in either direction, that depend on the context of other promises made by the agents.

An example helps to illustrate the difference between money already in circulation and borrowed money.

Example 37 (Borrowing money vs utility service). *Some analogies help to illustrate the semantics of money. Money is a kind of utility, like tap water, electricity, or Internet. The key difference between paying for money and the way we pay for utilities (service availability) is that we have to pay for use of money with money itself. Imagine if you paid for a car rental by bringing back bits of new car in payment: a wheel, a windshield...).*

We are taught to think that we get money for work (without asking where it came from). However, all money has to be traced back to someone's debt, even if that debt is eventually written off. When we buy a good, for ownership, we keep it and could in principle exchange it again for something else unless it perishes. We don't get all the money back, and certainly not with interest. When we buy a service utility, we keep getting replenished with new service (as if a rented office space were like a leaky tyre that got pumped up each month), but we don't get the money back. But in a loan, you don't: you get a fixed amount of air, and you are expected to give it back in better than the condition in which you received it. The legacy of treating money as an immutable thing has led to this conceptually difficult form of lending. The self-referential nature of paying for money makes it even harder to comprehend how to apply this analogy (it becomes not only non-linear, turtles all the way down.

Item Rental	Corresponding Rental
Electricity/network/office/car	Money loan (payment availability)
Give electricity/office space back	Repay loan
Monthly Bill for usage (covers costs then profit)	Interest on loan (pure profit)
Failure to pay (arrears)	Failure to pay monthly loan interest
Interest on arrears	Compound interest (loan + arrears)
Money for work (wages)	Someone pays into your account

Note that with money, repaying the loan is not the same as paying interest on the loan. The loan has the semantics of a rented item, while the interest has the semantics of a regular payment (in either direction, depending on whether the interest rate is positive or negative). When we rent an office space, we don't pay by giving back part of the office space each month, until we are finished with it or it is used up. Office space is not trivially additive, like money. In service payment plans, we can usually choose whether to pay for usage or pay a fixed subscription (all you can eat, up to some maximum quota):

Pay as you go (insert coins in meter)	Borrow money for fixed monthly interest
All you can eat subscription (max quota)	No analogue, except for the super rich

The 'all you can eat' option is not available for money, because it is still considered to be a representation for goods or services that are not to be returned. We return rented car or office space when no longer paying rent, and we return money when no longer paying interest, but we keep things bought forever. The main difference is that the money did not cost anything to make, so its loss has no cost to the lender, except a potential loss of trust in the eyes of regulators (who acts as society's trust police). Even though ultimately no one is truly disadvantaged by the loaning of money (other than by the arbitrary accounting in trust liability), our current system expects us to pay for borrowed money as if it brings continuously renewed utility, using money we obtain from other sources. The difference between loaned money and earned money is only that the earned money is already in circulation (having come from someone else's loan, elsewhere in the past). Money forgets this association, because it has no labels to remember where it came from. This makes us think it just exists, without having to be created like electricity.

6.4   ASYMMETRY BETWEEN MONEY AND DEBT

There is a semantic asymmetry between the debt created and the money of account created in such a creation event. Debts is a much more complex entity than money. This is philosophically interesting and practically important. It is philosophically interesting because it shows that our only mechanism for creating money does not create equal and opposite objects manifestations that sum to zero in all matters. In particular, it means that money is formally different from energy in physics (which it is often likened to), because it has additional semantics that are not even symmetrical between the deposits and debt (money and anti-money, in the analogy).

- Money is most fungible without memory. Mobile proxies with builtin ledgers, like BitCoin, have more in common with debt than pure money. However, whereas debt can be discharged and records can be let go over time, the same may not be true of cryptoledger technologies. This is could be a time bomb, or a slow death brought about by a poor design that doesn't take into account real world loss of semantic significance[34].

- Although we can use any money (from anywhere) to pay off debt, the payment of money into a ledger or bank account does not automatically discharge a debt, unless it promises specifically to do so.

- Traditionally, debt is accompanied by a positive rate of interest. If a loan holder fails to pay the agreed amount of interest, in the agreed amount of time, this introduces a new independent debt, which is independently accountable. In recent years, especially in Europe, negative interest rates have been employed to encourage lending and discourage hoarding. In either case, there are intentional semantics at work.

- The addition of service charges, rents, and other forms of levy, are not added to an original debt: they have different semantics. The cumulative amount of rent for the original debt plus any new debts, caused by failure to repay the original debt amount, may introduce compound interest. These derivative debts of the original one are semantically distinct.

- When a debt holder agent wants to discharge any one of these debts, it must specify which debt it is paying off (see figure 6.4, and compare it to figure 6.3).

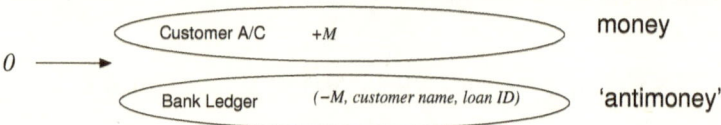

Figure 6.4: The reality of debt is more complicated than indicated in figure 6.3. Debts are labelled amounts, and the additional promises associated with the owner may remain even when the balance of payments is zero. Thus money and 'antimoney' do not cancel out as we would expect of the analogy with energy in physics. When companies are bankrupted, some information is cancelled out, but the knowledge of bankruptcy will also follow the agent affecting its future reputation.

- Loan payment contracts introduce complexity, with a multiplication of the payment promises, when certain promises are not kept in time, with new semantic labels. This is a non-linear, unstable process.

For some, debt is the crucial issue in money. Graeber, for example, has developed a detailed thesis of money, in which the primacy of debt obligation drives history with a similar degree of determinism as the differential equations of modern economics drive changes in economic measures today[Gra11]. Although his narrative is compelling, we can't help but feel that it presents a one sided view. People continue to associate in long term relationships not only because of debts to one another, but also because they can promise one another benefits. Nevertheless, we find his observation that 'freedom is slavery and slavery is freedom' to be a telling reminder that semantics are everything in the human realm. Some might prefer blind obedience to the responsibility of decision, while others would find the freedom to do as they please comes with an endless struggle to meet the expectations of others.

On a purely mechanical level, one cannot help but notice the similarity in the relationship between money and debt and electrons and holes in the theory of electrical conduction. Holes, like debt, are an absence of an electron (of currency), but they also carry more semantics: they are burdened with lattice properties from the environment in which current moves. This helps to illustrate that analogous situations also recognize and asymmetry between apparent opposites[35].

6.5 REMARKS ON CONSERVATION OF MONEY

It should be clear, from the foregoing, that money is not naturally conserved. This follows formally from the lack of any homogeneity in intent. However, we try to enforce sufficient homogeneity by convention or by regulation (see assumption 7), at least with respect to financial matters, to maintain the illusion of de facto money conservation. For without it, the concept of debt and ownership would be meaningless.

Conservation of some quantity does not mean we *cannot* create it or destroy it, only that we must account strictly for where it comes from and goes to, without that information getting corrupted. When we create an amount of money $+\mu$, banks also create a debt of $-\mu$ (a little like a matter-antimatter pair of virtual particles that are allowed to live for a finite time). Conservation is only a requirement of a local region. Conserved quantities can flow into and out of that region's boundary, leading to the appearance of sources and sinks of money. Again, as long as money is accounted for, there is no difficulty.

Money persists in circulation from banks that have bad loans (debts that will not be paid back), and which bankrupt for that reason. Then the money is circulating and can't be taken back from its current holders. Here, too, is an important asymmetry, because money of account cannot disappear or get lost in transactions—only by data loss.

Assumption 8 (Conservation of trust and money). *Society's intent is to account for money, without loss, as a matter of trust. The maintenance of trust is more important than the conservation of money.*

The loss of strict accounting can happen intentionally or by unintended data corruption. Assuring data integrity in all transactions is relatively expensive, for small amounts, but it is not optional as long as the policy of money accounting stands. Debt relief, write-offs, bankruptcies may be responsible for writing off large amounts of money, which remain in circulation. The consequences of this conservation are hard to assess, especially when money is stored in accounts with interest paid and levied. Would it be possible, for instance, to imagine a world in which there were such a high level of negative interest that most agents could not afford a positive income?

6.6   CLEARING MONEY OF ACCOUNT BETWEEN BANKS

When we make payments by cheque or card or transfer, nothing happens without the promise of the banks to honour the transactions, because the transactions only exist with the bounds of their ledgers. If banks hold property or commodities (gold bullion etc), then it may need to physically transfer some of those holdings to the destination bank, in case of certain transactions. Though, as we have shown, anything that can promise ownership can also be owned without necessarily having to hold it.

From the way in which money is created, we quickly identify two ways of transferring money between banks:

- Transferring the ownership of ownable assets, of agreed value.

- By agreeing to create and hold debts to one another (loans).

This raises the question of whether money can be owned by a bank (see section 6.8). In practice, banks create ownable money by inventing 'securities' and other technical forms of money.

Consider banks B_1 and B_2, with customer accounts A_1 and A_2, and ledgers L_1 and L_2 respectively. The issue of payment really boils down to whether money can be transferred directly from one account to another, from customer to customer, or whether the bank needs to be involved as an intermediary (delivery agent or courier).

- **Direct transfer account to account (Trusted Third Party)**

  Transfer account holder to account holder (peer to peer) assumes trust of the third party. This is easy within a single bank, or a set of banks regulated by an oversight agent.

$$A_1 \xrightarrow{+(+\text{transaction}(M,C))} A_2 \tag{6.23}$$

$$A_2 \xrightarrow{+(-\text{transaction}(M,C))} A_1 \tag{6.24}$$

  A_1 and A_2 are already agents, so we transfer ownership of a transfer amount by treating it as a subagent, and emitting and absorbing it, i.e. by transmitting a 'transaction' subagent. This is extremely simple, if the agents trust one another. This is simply the dual of the change of ownership sequence, with a payment in 'kind', where 'kind' happens to be money.

  The issue with this method is trust. Why would either side not simply increase their own balance and never decrease it? If agents are autonomous, why would they respect the transfer of information, without a third party to oversee? Third parties are straightforward, if trusted within a common scope. A single overseer, like a monetary authority or central bank, can examine all records and tally correlated payments to detect fraud.

The case is much more difficult if there is no single authority or jurisdiction, such as between foreign nations. This leads us to invent something like banks as go-betweens.

- **Transfer by proxy (bank to bank and beyond)**

For example, a transfer to another currency in a foreign country. Now the banks do not automatically trust each other by the calibration of banking standards according to their centrally regulated licence.

The answer to the first method of course is that we generally do need a third party to calibrate values to a common standard. Two agents could agree on the outcome of a single transaction, but may not treat others according to a common protocol of behaviour. This is the reason for introducing Trusted Third Parties[BB06b]. Banks play this role, hopefully under the eyes of government and judiciary; in the future there is no reason why computer programs or 'apps' could not perform this function much more directly and cheaply. The bank's role as a service provider is clear in all cases where money of account is involved. There is nothing physical to transfer.

With a third party involved, there are consequences for the complexity and reliability of the delivery (see chapter 11 [BB14a]). We can interpret this as a delivery through the intermediary of two banks:

$$A_1 \rightarrow B_1 \rightarrow B_2 \rightarrow A_2. \tag{6.25}$$

Treating this as an assisted promise (assisted by two banks). The payer promises a transfer conditionally on the banks making a transfer by proxy:

$$A_1 \xrightarrow{+\text{transaction}(M,C)|\text{bank transfer}(A_2,B_2)} A_2 \tag{6.26}$$

$$A_2 \xrightarrow{-\text{transaction}(M,C)|\text{bank transfer}(A_2,B_2)} A_1 \tag{6.27}$$

$$\tag{6.28}$$

The sender's bank agrees to pick up the transaction and deliver it as far as the second bank[36]. . If we simplify the notation slightly, one can see the symmetry of promises (see 11.3.3 of [BB14a]). Let $T \equiv \text{transaction}(M, C)$, and X_1 be the conditional promise to perform the bank transfer from B_1 to B_2. Then let X_2 be the promise to deliver from B_2 to A_2:

$$\text{schematically}: \text{``}A_1 \xrightarrow{T|X_1,X_2} B_1 \xrightarrow{X_1|T,X_2} B_2 \xrightarrow{X_2|T,X_1} A_2'' \tag{6.29}$$

Then, in the notation of [BB14a], we must promise:

$$A_1 \xrightarrow{\pm T(X_1(X_2))} A_2 \tag{6.30}$$

$$A_1 \xrightarrow{\pm T} B_1 \tag{6.31}$$

$$B_1 \xrightarrow{\pm X_1(X_2)} A_1 \tag{6.32}$$

$$B_1 \xrightarrow{\pm X_1(T) \wedge X_2} A_2 \tag{6.33}$$

$$B_1 \xrightarrow{\pm X_1(T)} B_2 \tag{6.34}$$

$$B_2 \xrightarrow{\pm X_2} B_1 \tag{6.35}$$

$$B_2 \xrightarrow{\pm X_2(X_1(T))} A_1 \tag{6.36}$$

Complete awareness of these promises may seem excessive compared to what we do in daily life. But we should point out that we fail to verify most of these promises in daily life because it costs too much, and we prefer to trust instead. However, in the case of building up such trust, when dealing with new intermediaries and, third party bank-like agents, verification of the strict keeping of all of these promises is part of securing fair transactional integrity.

Understanding this algebra opens up the possibility of opening up the role of trusted intermediary to more than banks. Indeed, this suggests that banks have a finite time to live, as service providers, in the information age. Interest policies will price them out of the market.

6.7   EXCHANGE OF CURRENCIES, AND FOREIGN TRANSFERS

Money only works as a network of agents agree to accept it. The virtual boundary in which money is accepted forms a region both of semantic applicability and of trust in monetary authorities.

Definition 58 (Common currency region). *A superagent R formed from all the member agents $A_i \in R$ that accept a given currency in exchange.*

$$(A_i \in R) \xrightarrow{-\mu_{currency}} (A_j \in R) \tag{6.37}$$

Within a common currency region, money acts as a lingua franca for promising between agents [Bur14, Bur15, Bur16b].

We are all familiar with the buying of foreign cash notes. This requires a physical transfer of mobile money. More interesting is how large amounts of money can be moved.

The transfer of monies, with currency conversion is performed by the sale of one currency for another. It cannot easily be carried out directly by account holders using money of account, because that would requires a bank account in the currency. A bank needs to intervene to handle the accounting in a user transparent fashion: the conversion can be made as a coordinated creation of local bank money, using assets to cover an averaged stream of equivalent exchanges.

Example 38. *A domestic bank in region A, with currency A, sells an amount of currency μ_A (probably aggregated over multiple smaller transactions into a large amount, for efficiency) to a foreign bank in region B, at an agreed rate, for amount μ_B of currency B. The amount μ_A is credited to the bank in currency A's region. This money is created in payment for an asset owned by the bank in region B, worth μ_B, which is linked to the sale of currency A. No money actually has to leave its protective boundary. Virtual assets can be bought and sold to compensate for movements back and forth.*

These assets may be held in the banks of the foreign country, so that the currencies never truly leave their country of origins. Effectively foreign countries would have an account at a foreign bank, and an agreement between banks to hold one another's currency assets on their ledgers, but no country can create the other's currency. The regulation of these amounts is somewhat ambiguous, and cannot be regulated in a foreign territory. Thus, it is important that international treaties play a role in money exchange, and the assets are tracked carefully.

Definition 59 (Foreign exchange (Forex)). *An agent who trades in different currencies by holding assets and reserves of different currencies in various forms.*

Example 39. *Global retailers who accept money in multiple currencies often allow credit card users to select the currency of their choice, usually to pay in the native currency of their account, when abroad. This avoids the costs incurred from exchanging the currency, through a third party, and the transferred amount remains in the currency of choice for the retailer, all assuming one accepts their rate of exchange. This may avoid fees exacted for buying and selling both at the point of sale, and in the future for the retailer. It is an innovation that allows a single credit card acceptance point to pay into multiple accounts, selected by currency type.*

There are several issues here that warrant a more complete description.

6.8 WHO OWNS MONEY?

Pure money is not ownable: since it only makes a single promise $+\mu$. It can be held and owned implicitly by being encapsulated by something that can make promises of ownership. Money *proxies* can be owned, if they make sufficient promises to be consistent with that view. Coins and notes cannot labelled to reflect a change of owner (at least in any physical currency we know of), so they can normally only be held (see table 5.1)[37]. Coins are effectively labelled with the mint that produced them, so the coins remain the property of the Mint. Bank notes are sometimes numbered: this makes them distinguishable, and allows a ledger to assign an owner to a specific note number, though we would be surprised if this was ever done, given that the difference between notes is negligible. Other kinds of promissory notes, such as 'IOU (I owe you)', are labelled and thus are owned. Money of account is owned by the account holder, by virtue of never leaving the container of a bank account, which is labelled. Thus they are effectively labelled (section 4.7). Coins and notes could be argued to still be the property of the bank or mint that made them, as they do often have a label from the producer, and they are also ultimately responsible for what happens to them. Other proxies might remain owned by the banks or the minting source, or be released into 'freedom'.

Physical notes can be labelled with watermarks and digital signatures, as can transactions of many cryptocurrencies like BitCoin. However, ledger money is owned only by virtue of existing entirely within the boundary of the ledger, and having no existence beyond it. This depends on how much ancillary documentation is kept in the ledger. One cannot change ownership of a BitCoin ledger entry, one can only create or destroy. Thus transactions involving direct transfers of bank account money are manipulations on independent ledgers of the different banks. In this way, every bank operates an entirely separate currency, i.e. effectively maintains its own autonomous money. The amounts in transactions are ownable, but the money itself not.

Blockchain cryptocurrencies have the technology to promise ownership on transactions, and thus ownership of amounts can be changed. As with bank money, it is difficult to argue that the coins themselves can be owned since they have no semantics other than an amount. The semantics lie in the ledger of transactions. In general, there is not much benefit to owning a proxy unless you are afraid of losing it (you don't trust agents around you), or are trying to rent it out as a service (bank charges apply to the account, so in principle all money of account). Agents may own pure money only by containing it, since money is only a single indistinguishable promise ($M \xrightarrow{+\mu} *$) with no way of representing ownership. Monetary authorities (banks, central banks, states, Starbucks, Air France, etc) own the proxies (including data transactions). Banks can get around the ownership of money by inventing forms of money, called securities and bonds (bindings), which add additional semantics to the proxies.

6.9 ACQUIRING MONEY

How we acquire money is a different question than where it comes from. Moreover, acquiring money is different from acquiring wealth, because money is invariant, whereas value is relative.

6.9.1 PERMANENT TRANSFER

There are several routes by which agents can acquire money:

- They can create it themselves. Unless this is authorized and regulated, it could render money useless as a means for limiting the free access to goods and services, since it would be the same as making everything free.

- Gifts and inheritances. We might be given money proxies as a gift.

- Selling goods for payment. We can exchange things for money.

- Selling services (including labour) for payment. We can exchange action or outcome for money.

- Borrowing or loaning.

6.9.2 TEMPORARY BORROWING

When agents need access to a reservoir of savings (money) they don't have, in order to overcome some obstacle, they need to borrow from some Trusted Third Party. Previously, money could only be borrowed from family or money lenders. Today, we have a network of financial services, offering commercial terms for borrowing money, with different incentives. We find three basic mechanisms for routing money to overcome obstacles.

1. Private borrowing (loan, mortgage, or credit payments), e.g. from a bank, which is authorized to create new money.

2. Taxation and redistribution of money by from all citizens, channelled into welfare, grants, or public concessions.

3. Selling shares in an enterprise for an expected later profit (investment or gambling).

4. Crowd-funding, borrowing or investment from distributed social groups savings. This includes collectives, such as local communities creating funds for eventualities.

These are sometimes called variously capitalist and socialist methods; but, politics aside, they are just network architectures for redistributing money traffic, and different boundary conditions in time. In all cases, except creation of bank money, someone has to save up for such eventualities in advance. This kind of planning avoids the payment of positive interest; it is an insurance model. By saving up for a Thing, a group takes out a kind of 'lack of Thing insurance'. Savings can have the opposite effect, if they cannot be used. If money is locked up in hoards, it does no one any good (to use the network analogy, it's like keeping a field full of cars for private use when transport is in short supply).

The conservation of money suggests that lending money would be an altruistic act: if an agent lends money to other agents, then that money becomes unavailable for personal use, and poses a possible risk. Specifically, an agent might need the money to overcome an obstacle of its own. Financial innovation thus led banks to invent the money for loans, so that no one was disadvantaged by a loss of money availability in practice. In that model, borrowing money increases the total supply of money (or the 'bandwidth' of the monetary network).

Example 40. *In traffic management, another form of transport network, routing management is needed. We can rent cars (like borrowing money), but since we don't pay in smaller cars, the semantics of owing rent and borrowing cars are distinct (the interest cannot be cumulative). If too much traffic goes in only one direction, all cars end up in one town and transportation breaks down for the rest of the towns, until such a time as someone in the other town wants to invest time in the other towns, or the other towns can borrow cars. Cars could be also be borrowed directly from neighbours (crowd-funding) on a friendly basis, or could be provided by the state (buses).*

If one discounts borrowing from banks, who are authorized to create new money, subject to a few restrictions, then all borrowing would have to come from the accumulated savings or assets of individual agents, or by delaying payment in time (which is like abstractly borrowing from those to whom we have already promised money). But the total amount that could be saved can never be more than a fraction of the existing supply of money in society. What if that fraction is not enough to overcome the obstacles faced? That is one reason why new money is created during borrowing, but it omits the most salient point in capitalist borrowing, which is the historical practice of charging of rent on debt, so so-called (positive) 'interest'.

6.10   THE COSTS OF MONEY

Money has a deliberate cost associated with it. It may seem paradoxical that acquiring or holding money should have a cost, knowing that it costs nothing to create money, but the cost amounts to a mixture of administrative overheads and a policy decision to limit its use. If access to money were free, there is a concern that we might create too much of it and hoard it, just as we accumulate network data quotas from our network carriers. The difference between personal data networking and money is that one cannot actually use network capacity all in one go, because the infrastructure throttles the rate at which we can use it, whereas there is no obvious limit to how much could be spent in a day, unless banks place limits on this.

The expression 'cost of money' has a specific meaning in contemporary finance, which we mention here because it pertains to borrowing, though a fuller discussion of interest must wait a discussion of prices in section 7.14. We begin with the notion of a cost, in the general case:

Definition 60 (Cost). *A negative monetary dependency associated with the keeping of a conditional promise π, promised to the third party agent who promises to fulfill the condition.*

Example 41. *A retailer promises apples at a certain price, conditionally on being able to buy the apples from a supplier. The acquisition of the apples imposes a cost on the retailer, to be paid to the supplier.*

There are several ways in which costs are imposed on customers, in relation to money:

1. When we enter into a tenancy for the holding of a bank account.

   Practices vary widely today. At one time, banks used to charge a fee for holding money (accepting a risk of storing valuable stuff). Today, it is more common that we are paid to place our deposits so that banks can earn money by investing them. What was a cost became a business model. Bank licences depend on maintaining a balance of deposits at a certain level, which restricts their ability to create new money. The charged price is accepted as a cost of doing business.

2. Banks may have costs associated with creating physical proxies for money (minting costs of coins and notes, printing of cheques, giros, etc). Today, this includes power bills for computational resources.

3. Transacting transfer fees, include the cost of human labour or computational infrastructure. These costs are approaching zero today.

4. Tenancy rent, a service fee for holding an account.

5. Interest on loans and mortgages.

The origin of the idea of charging interest in worthy of a book in its own right. Not all banks charge interest. In Islamic finance, for instance, the charging of interest is forbidden[BM11, Ber15].

Without central law, banks could very easily defraud society with impunity, and trust in banks as institutions might quickly collapse[38]. Banks hold the licence to create money on our behalf. More importantly, banks are allowed to charge borrowers rent proportional to their amount of outstanding debt. Clearly banks do not have to, from a technical perspective, but their business model is based in this rent. In proximate terms, banks argue that they take a risk by not having a larger buffer of money at hand than they they have after lending. This allows them to seek remuneration for the service they offer. In finance, the concept of a cost of money is related to this interest imposed by a central bank:

Definition 61 (Return or Cost of money). *A negative amount of money, imposed by a lender, equal to the profit which could have been earned on a given base amount of money if it were invested in government bonds, or if the expected outcome of investment in government bonds is negative an equal positive amount imposed by the borrower on the lender.*

We return to discuss this further in the context of interest, in section 7.14.

6.11   THE ENTROPY OF NETWORK MONEY

The concept of entropy (originating in physics and information theory) is renowned for its subtlety and even for its popular abuse. Entropy is simply a measure of the extent to which something is distributed across a set of possible outcomes. The subtleties arise in the interpretation of that basic idea. In physics we are familiar with the popular notion if thermodynamics entropy being an estimate of the extent to which energy is distributed in such a way that it cannot be used to do useful work (often called disorder).

Entropy is effectively a measure of how distinguishable different parts of a system are from one another, in terms of their promises. If every part of a system is identical, it is hard to imagine how anything further of interest could happen. Activity in any kind of system requires there to be inhomogeneities in the distribution of stuff. In economics, stuff means goods, services, and money. From an information perspective, an economy only makes sense in a state of disequilibrium.

Because entropy describes how 'well mixed' things are, it is also an implicit measure of how expensive it would be to separate them, or recycle the raw elements from the mixture. Entropy is thus related to the cost of extracting information. It is formally a factor group:

$$\text{Extensive entropy} = W \;=\; \log\left(\frac{\text{Distinguishable states}}{\text{Equivalences}}\right) \tag{6.38}$$

$$W \;=\; -N\sum_{i=1}^{C} \log p_i \log p_i, \tag{6.39}$$

where p_i may be interpreted as the normalized frequency of a system transaction being in a particular state i of an alphabet of C different states, over N transactions. If information is never lost or muddled, the entropy is low and the information cost of maintaining it is all up front, so that there is no recovery cost. If, on the other hand, information is muddled, merged, or labels are removed so that we can no longer tell the difference between different states $C \to 1$, then the amount of entropy is large, so it costs less (but is also less useful). This is easy to see by setting $p_i = 1$ implying $p_{j \neq i} = 0$, leaving $S = 0$. Any other case has $S > 0$.

In a promise theoretic system, the distinguishable states are represented by promises (made by agents). The distinguishable states of an economic system are:

- Different agents (represented by bank accounts or cash holders) or identity promises.

- Different payments of money held by these, or monetary and price promises.

If amounts are all lumped into a featureless sum total, they can no longer be distinguished, information is lost, and entropy increases. This happens when we take away promises, like failing to keep records of the source of the transactions. Clearly cash (coinage and notes) retain very little information about their exploits, and thus lead to high monetary entropy. BitCoin remembers a lot in its ledgers, and has low entropy, but high cost to maintain. Debts carry higher information than money, and thus are more expensive to document than money deposits. Note that these costs are transaction costs, and are unrelated to the costs incurred by charging of interest on loans, etc.

The same entropic principle applies to the infrastructure of money, both for transactional transfers and for prices (kinetic and potential money). The network complexity of interactions has to be borne by the agents themselves, so standardization of exchange by a locally central hub, like a bank, can reduce the objective costs experienced by the agents.

Example 42. *When all transactions and exchanges are specifically labelled between pairs of identifiable agents, there is a cost of trusting and dealing with $O(N^2)$ relationship states between the agents involved. When payments are routed through a bank (a third party), then there is only a single kind of payment and currency and the entropy is* *low (C = 1) for all agents except the bank (which has to maintain $O(N)$). Thus the cost of routing transfers through a third party costs a lot less for each agent than having to keep information about every potential trading partner. A single central banks (or any locally central structure) 'coarse grains' the detailed network of trade happening within its ledger, and presents it as a black box, lowering the semantic costs to agents outside it. If more banks and currencies are added, at the scale of inter-bank trades those benefits dwindle, as one recreates the same network complexity at a larger scale. If all agents were their own bank, the benefits of local centralization would be gone.*

Although local centralization can reduce the objective (dynamical) costs to agents, this is not the end of the story. In a capitalist system, the bank might impose its own costs of interaction on its clients by setting a price for its services. This would be an intentional social cost, imposed as a levy, which might well not exist for direct trade. Social semantics might therefore override dynamical favourability, meaning that agents are unable to take advantage of the cost savings implied by the entropy of natural network configurations.

Entropy highlights one thing however, which is the potential cost benefit of eliminating semantics. The classic information dilemma, as applied to money, is that the fewer semantic labels money has, the more fungible and acceptable it is (labels cannot be used to discriminate or poison its neutrality): the cheaper it is to accept and pass along, but the more information is lost, with potential consequences for trust. Indeed, entropy is how money is laundered to shed its origins. On the other hand, what if traceable semantics could actually solve a larger problem, on a different functional scale, invisible on the scale of the network. Then the total systemic cost might be offset by an even greater profit within a certain region of the total network, due to the semantics of prices and transactional flows. There is therefore a great temptation for agents whose motive is profit to exploit these informational characteristics to divert money[39].

While the money emanating from a loan is traditionally entirely unencumbered by labels, and is indistinguishable from any other money, the debt, which is created along side it, retains its labels, so that they must be repaid by the specific borrower according to the specific terms[40]. This is an opportunity from which banks can profit. Debts could, of course, not be practically repaid if money were not created indistinguishable from other money (with already high entropy). For, if loan money did not forget its origins, the exact transactions would all have to be rounded up and paid back, serial number by serial number, to settle the debt like a precise jigsaw puzzle. On the other hand, if debt carried high entropy (no labels) one would not be able to hold the individual borrowers accountable for their borrowing. Entropy shows us that a lack of trust in an economy is a hindrance to doing business.

6.12 RETURN ON INVESTMENT, SURPLUS, AND CASHFLOW QUEUES

It is practically an axiom of capitalist economics that agents should seek to make a profit or surplus of money. The point of this, historically, was independent of capitalism: it was to enable agents to advance in some way. By developing time-saving technologies for creating a surplus, we spend less time on subsistence to invest in creating even better innovations, in an upward spiral. This is not without controversy[Suz17], but it seems to fit the scaling of larger societies[BLH+07]. Profit is thus desired in order to invest back into society, buying *time* to support long-term development, growth, and better living standards[Dia97, Har11]. More recently, shareholder capitalism has transformed this tradition into a profit and rent-seeking strategy for maximization of shareholder profit. This has further sparked controversy about what the purpose of the global economy actually is. In this context, it is sufficient for us to assume that the gathering of surplus is an almost universal empirical behaviour in contemporary society.

Assumption 9 (Profit and ROI). *All agents attempt to keep the dimensionless monetary ratio of sales/purchases greater than 1, over some identifiable timescale, maintaining a surplus in their balance of payments, for the purpose of future advancement.*

The timescale is clearly of importance, as it says something about the relative rates of stochastic sales. In business, quarterly earnings dominate boardroom discussions; for smaller businesses, weekly or monthly cashflow marks the breadline. For corporations and governments, debts can be kept for decades without ill effect, gambling on future returns. Borrowing and savings buffers play an obvious role in this ability to keep cashflow balanced.

There is a tendency to think of cashflow in terms of a steady state dynamical equilibrium, but this is misleading. A steady state picture of stochastic events is much like a 'queue', in the sense of queueing theory[Kle76]: a statistical aggregation of individual events happening with a certain probability. If money comes in with an arrival rate of λ_q, and goes out with a service rate of μ_q, then the queue can balance on timescales $\delta t \gg \lambda_q^{-1} \gg \mu_q - 1$. The dimensionless ratio $\rho = \lambda_q / \mu_q$ is the 'traffic intensity', and signifies a critical failure as $\rho \to 1$. The analogy to money as a form of network traffic is appropriate, and we should expect the same behaviour. This tells us that money flow is potentially unstable to accumulations of money, and needs continuous new input of money, or loss of recipients, to support the idea of profit.

Dimensional analysis can help us to see how the amount of money in circulation needs to increase to support the notion of continuous profit by all. All scale free invariants must be expressed in dimensionless variables. Given that some monies may be in savings (queued up), or in transit, an excess amount of queued savings may work to the advantage of continuity or may prevent continuity depending on the network arrangement. What one immediately expects from simple dimensional analysis, and a dynamical similarity with queues[Bur13b], is that monetary flow will have locally critical behaviour based on the following scales: money rates, times, and money:

- The rate at which agents can pay $\lambda_q \sim [\mu/t]$.

- The rate at which agents can sell $\mu_q \sim [\mu/t]$.

- The amount of savings each agent holds $\mu_{savings} \sim [\mu]$.

- The timescales embedded in contract semantics $[t]$.

Square brackets denote engineering dimensions or type of units.

CHAPTER 7

BUYING, SELLING, AND PAYMENT

Money operates in a world of interacting agents. Like any networking technology, it needs to 'flow' to fulfill its purpose. We shall not assume that this flow is smooth or differentiable, as it might be on very large scales, but analyze its discrete and stochastic nature on the small scale, in keeping with earlier assumptions and the model of queues noted in the previous section. We have already noted that money provides a kind of common standard networking infrastructure that pushes differences in to the edges of that network, where individual and subjective judgements can be localized within trading agents. In this section, we consider some simple properties of money as a stochastic system.

7.1 TRADE

In common parlance trading often implies a direct exchange of items for other items. A cursory glance for definitions in the literature shows that trade is now almost universally defined in terms of buying and selling (implying the use of money). Indeed, Graeber believes that money has been integral to trade throughout the history of civilization[Gra11]. We shall try to keep the notion of barter trade (without the intermediary of money) distinct from buying and selling, since it is a possible mechanism, whatever its prevalence, without taking a position on their relative importance.

Definition 62 (Trade and exchange). *Trade is a bilateral exchange of things held or owned between two parties. Trade may be accomplished virtually or physically:*

1. *Physical: The emission of an agent representing a good by one (sender) agent S, and subsequent absorption by a recipient R.*

$$S \xrightarrow{+g_1} R \qquad (7.1)$$

$$R \xrightarrow{-g_1} S \qquad (7.2)$$

$$S \xrightarrow{-g_2} R \qquad (7.3)$$

$$R \xrightarrow{+g_2} S. \qquad (7.4)$$

where g_i is an agent of exchange[Bur15]; g_i may be a good or a monetary token.

2. *Virtual or service: The promise of a service by one agent and the acceptance and use of it by another.*

$$S \xrightarrow{+s_1} R \qquad (7.5)$$

$$R \xrightarrow{-s_1} S \qquad (7.6)$$

$$S \xrightarrow{+s_2} R \qquad (7.7)$$

$$R \xrightarrow{-s_2} S. \qquad (7.8)$$

3. *Mixed: Trade of a service for a physical renumeration.*

$$S \xrightarrow{+s_1} R \qquad (7.9)$$

$$R \xrightarrow{-s_1} S \qquad (7.10)$$

$$S \xrightarrow{-g_2} R \qquad (7.11)$$

$$R \xrightarrow{+g_2} S. \qquad (7.12)$$

Since money may be considered either a good or a service, these cases also account for exchanges that include money.

At this stage, we do not need to specify any implied timescale or notion of simultaneity. This will come about by considering the semantics of trade balance. Before proceeding, it is worth noting that this kind of exchange, without money, is equivalent to a trading of gifts.

Definition 63 (Gift). *A voluntary change in the ownership of a thing* T, *from an agent* S *to* R, *without the expectation of remuneration, payment, or exchange. Gifts may be offered as promises,*

$$S \xrightarrow{+T} R \tag{7.13}$$

$$R \xrightarrow{-T} S \tag{7.14}$$

or imposed as impositions:

$$S \xrightarrow{+T} \blacksquare R. \tag{7.15}$$

In the case where gifts are impositions, there is often an attempt to implant an obligation to repay the gift somehow, to maintain a sense of honour. This game strategy can be considered a backhanded form of 'attack' in the Promise Theory sense (an attempt to induce cooperation without prior warning).

Example 43. *Companies send free samples of goods to home owners, with an automatic subscription in the small print. The recipients often have to buy more or return the gift at some cost in order to cancel the subscription.*

This kind of strategy bears some measure of subterfuge; indeed, in history, it has been argued that money and gifts were often connected with acts of violence in a long standing and sinister relationship[Gra11].

Lemma 15 (Trading and gifts). *Trade without money is equivalent to a mutual offering of gifts.*

This follows immediately from the definitions in (7.14) and (7.1-7.4). From the symmetry, one could also ask the question, if is equivalent to mutual stealing or extortion? We assume not, as these would be represented as impositions. Gifts are not thought usually of as being imposed on the recipient, though this is strictly an assessment of the recipient.

Trade is usually held in some kind of balance, as a measure of trust and responsibility.

Definition 64 (Trade balance). *A state characterizing an equilibrium within a local network of trade, measured over an agreed timescale. Acceptable ranges for incoming and outgoing exchanges, at each agent, are agreed by all agents in the network, including the amounts and kinds of things promised, and the agreed time interval for completion, together with a specification of how to asses when all promises have been kept to their mutual satisfaction.*

The simplest case would be to define trade balance as something between pairs of agents, but this does not scale (its scale is pinned to particular agents). Also, it does not take

into account the network aspect of trade, or policy about what we mean by acceptable
or fair distribution. Here we might note the problem of trade balance as a concept: it
requires *agreement*, and defies a simple definition in purely quantitative terms[41]. The
role of semantics is never clearer than when political context alters agents' assessments.

If the agents promise to accept the terms of a 'fair' or 'acceptable' trade, i.e. that both
are satisfied with the utility they acquire, then they discharge any possible obligations
in the future (see 8.4 in [BB14a]). The following promise proposals form such an
agreement.

Definition 65 (Acceptable trade). *When both agents S and R promise to accept these
(e.g. by signing), then both sides have promised acceptable trade*

$$S \xrightarrow{v_S(+g_1) \le v_S(-g_2)} R \qquad\qquad (7.16)$$

$$R \xrightarrow{v_R(+g_2) \le v_R(-g_1)} S \qquad\qquad (7.17)$$

*i.e. S promises that its valuation of goods offered is less than or equal to its valuation of
what it accepts, and vice versa. So each agent feels it got a 'good deal'.*

When both agents S and R promise to accept these (e.g. by signing their affirmation),
then both sides have promised fair trade (see [BB14a]). This represents a two-person
game, and has a Nash equilibrium solution[BB14a, Bur13b].

Notice that no comparisons need to be made outside of the pair of agents S, R. Thus
there need be no absolute calibrated value for exchanges.

Definition 66 (Fair trade). *Fairness may also be assessed by a neutral third party T.
When both agents S and R promise to accept the promises (e.g. by signing), then both
sides have promised fair trade*

$$S \xrightarrow{v_T(+g_1) \le v_T(-g_2)} R \qquad\qquad (7.18)$$

$$R \xrightarrow{v_T(+g_2) \le v_T(-g_1)} S \qquad\qquad (7.19)$$

i.e. the same valuations are now made by T, as the neutral arbiter of fairness.

Trading is a system with few semantics, and little predictability, somewhat like a kind
of weather system of interactions whose stability is hard to assess. The question of
whether this closed network of transactions is predictable or stable, and can maintain a
functional distribution of wealth across the network (representing society) is effectively
an eigenvalue problem at each epoch of time [BBCEM10], but not necessarily one we
can compute. If the promises to behave according to this system are not kept, anything
could happen.

7.2 BUYING AND SELLING

Buying and selling are about intended exchange, in which the remuneration for things is paid in money. This statement is not entirely free of ambiguity, since even the simplest form of accounting could be considered money (e.g. ledger money of account), and any kind of table of equivalences in terms of direct goods can be interpreted as a form of money. Buying and selling are also voluntary interactions, freely entered into by both sides. Graeber argues that, outside the scope of the capitalist economy, some transactions are driven by debt and obligation rather than autonomous desire[Gra11]. We shall not consider these cases here.

When money is employed, through some proxy technology, then once we assign the role of buyer and seller (as opposed to neutral trader), the direction in which money flows selects a polarity to trade. Buying is the offer of proxy money in remuneration, and selling is the trade of goods for proxy money. This parlance now seems universal.

Definition 67 (Buyer or customer). *A role for an agent B, characterized by the intent to buy a thing T, equivalent to a promise that may be written:*

$$B \xrightarrow{-T} * \qquad (7.20)$$

Definition 68 (Seller or vendor). *An role for an agent S, characterized by the intent to sell a thing T, equivalent to a promise that may be written:*

$$S \xrightarrow{+T} * \qquad (7.21)$$

Although it is tempting to imagine that buying and selling happen as a simple synchronous handshake transaction, in modern economy trade happens in an *asynchronous* manner, and money is the glue that allows us to play with time and promise the terms and conditions agreed by the parties.

Assumption 10 (Ownership, buying and selling). *An agent cannot be bought or sold unless it is owned i.e. it is property. Once bought it becomes the property of another owner.*

Goods or services manufactured by an agent may be considered the property of the manufacturer as long as the resources used to make the thing were already owned by the agent.

The polarity of buying and selling for money is clear cut. All the semantic content of the transaction is attached to the things purchased, since money (as we define it) has no intrinsic semantics relative to trade. Ideal money is an invariant, by design. This state of affairs might not persist into the future, however, as we begin to entertain microcurrencies with specialized semantics, e.g. the use of particular private currencies to signify loyalty.

Things we buy represent influences that tend to attract our money away from us, and the offer of services (labour) tend to attract money back towards us again. In the industrial age, it was a seller's economy, and goods were mass produced as commodities for cost efficiency. Buyers took what they could get. In the information economy it is a buyers economy, and sellers have to manufacture cheaply whatever buyers want, as profit margins shrink to nothing[Tof70, Tof80, Rif15, Sea12].

7.3   PAYMENT

Payment, in its most general sense, is the act of giving up of one thing in remuneration for another thing[42]. Whether trading directly, goods for goods, or services for services, or using money, an agent in the role of buyer can refer to the keeping of some promise as payment for its receipt of the other's promise kept. Thus, keeping within the framework of promises, we can define (see 7.4.1 in [BB14a]):

Definition 69 (Payment). *The keeping of a promise of to pay an amount $\mu(P)$, where P is the agreed price, by a buyer B, in return for the acceptance of a promise of R, kept by seller S, is a payment of P from buyer B to seller S :*

$$B \xrightarrow{+\mu(P)|R} S \qquad\qquad (7.22)$$

$$B \xrightarrow{-R} S \qquad\qquad (7.23)$$

The semantics of payment are *conditional* on receipt of the 'goods' R, and thus we use a conditional promise, apparently indicating a prerequisite order of events. Note however, that the order is not inevitable, if the the counter promise to provide R is also conditional:

$$S \xrightarrow{+R|\mu(P)} B \qquad\qquad (7.24)$$

$$S \xrightarrow{-\mu(P)} B. \qquad\qquad (7.25)$$

In other words, if the offer of the thing being sold is also made conditional on payment, then this creates a stalemate: who moves first? This symmetry simply has to be broken by one of the parties, in order to determine the sequence, as an act of faith or trust[Axe97, Axe84].

One might also assume, conservatively, that payment and goods are positive quantities. However, when semantics and money mix, payment becomes both subtle and intricate. Usually, one expects an amount of money $\mu > 0$ to be strictly positive in exchange for a good or service. However, some goods may be given away, and, in some cases, goods may even be given away with a bonus, where $X < 0$ as promotions.

Definition 70 (Payment in kind). *An expression still used to indicate a trade of favours, goods, or things that are not moneylike, in return for something that appears to carry a monetary measure.*

Payment in kind is thus a form of barter, usually used today to avoid monetary information being recorded, measured, and taxed.

Payment involves a protocol (applying (**Ax3**)):

Definition 71 (Payment (with money)). *A measure of the money to be transferred from sender to receiver during a change of ownership. The promise of an initial asking price is optional:*

$$S \xrightarrow{+right\ to\ purchase\ for\ \mu_X} B \quad (display\ asking\ price\ unconditionally) \quad (7.26)$$

$$B \xrightarrow{-right\ to\ purchase\ for\ \mu_X} S \quad (acknowledgment\ of\ price) \quad (7.27)$$

It could be omitted by moving directly to an offer to buy. The minimal, necessary and sufficient promises for payment are the following:

$$B \xrightarrow{+money\ amount\ \mu_Y} S \quad (make\ offer) \quad (7.28)$$

$$S \xrightarrow{-money\ amount\ \mu_Y} B \quad (acknowledge\ offer) \quad (7.29)$$

$$S \xrightarrow{+exchange\ goods\ for\ price\ \mu_Y\ |\ money\ for\ \mu_Y} B \quad (acknowledgment\ of\ price) \quad (7.30)$$

$$B \xrightarrow{-exchange\ goods} S \quad (accept\ goods\ unconditionally) \quad (7.31)$$

where each + promise is accepted with a - promise, according to the principle of autonomy, or local information.

We move to assuming payment in money here. This is almost universally assumed today. Why can explain this as follows:

Lemma 16 (Money equalizes opportunity). *Within a single currency region, when all agents in a network promise payment in the single money currency, then all offers can be received without error.*

This follows essentially from Shannon's error coding theorem. Within a single currency region, if the exchange payment $P_\mu \in \mathcal{P}$ is a mapping into the same monetary currency μ, then all agents can assign a price within a single alphabet, and all agents have the same opportunity to accept.

Example 44. *Money allows a coding can be made without loss of information. In the following transactions, the payments are proposed using alphabets that a not congruent.*

There is no unambiguous mapping for currency conversion.

$$\Sigma_1 \quad \rightarrow \quad \Sigma_2 \tag{7.32}$$

$$\{Goat, Pig, ...\} \quad \rightarrow \quad \mu \tag{7.33}$$

$$\{Goat, Pig, ...\} \quad \rightarrow \quad \{Bag\ of\ wheat, Keg\ of\ beer, ...\} \tag{7.34}$$

$$\tag{7.35}$$

7.4   PRICE

The language of trade and commerce is the communication of prices. Price is how intent enters a monetary network. Prices act as a distributed collection of promises, at the edge of the network, that signal an intent to sell (or by complementarity of + and -, an intent to buy). It is now most common to express payment in money[43], but a price could also be asked in any form, such as a swap of goods or services. For reasons of taxation (and redistribution), the pricing of goods by means of an exchange is generally forbidden. This contributes to making money semantically more important than that of a proxy for exchange.

7.5   THE NATURE OF PRICE

The most basic question about price, from a Promise Theory perspective must be to decide whether prices are promises, impositions, or assessments. An assessment is a private local matter, which cannot be passed on to another agent without a promise or imposition, so whatever role assessments play in deciding the asking price, it can only become effective by communication as a promise or an imposition. Thus, we can relate these as follows:

Definition 72 (Assessed price). *A price $P(T)$ may be offered or imposed from a seller S to an agent B, either unconditionally or conditionally on a valuation. If conditionally ((**Ax3**) and (**Ax4**)):*

$$S \xrightarrow{\ P(T)\ |\ v_S(T), v_B(T)\ } B \tag{7.36}$$

$$S \xrightarrow{\ P(T)\ |\ v_S(T), v_B(T)\ }\blacksquare\ B \tag{7.37}$$

then the assessments of valuation $v_S(T)$ and $v_B(T)$ are also a part of the promise, then it also becomes known to B and B will expect some (bounded) rationale for the relationship between the two.

On the other hand, if the price is merely stated unconditionally:

$$S \xrightarrow{P(T)} B \qquad (7.38)$$

$$S \xrightarrow{P(T)}_{\blacksquare} B \qquad (7.39)$$

then it may be considered *ad hoc*, and B ultimately bases its acceptance on its assessment of trust of S.

There may be circumstances in which a price can be imposed. This assumes that an agent is somehow obliged to accept the imposition. The semantics of such a state are not easy to generalize, e.g. there might be multiple suppliers or price fixing.

Lemma 17 (Acceptance of an imposed price). *An agent A must accept an imposed price iff:*

1. *A needs T and S's specific terms of sale Π_X (delivery promise, etc).*

2. *A has no alternative to S and depends on T, regardless of the terms of sale (monopoly).*

For the first case, T is a dependency, and so is X, so acceptance is conditional on some additional bundle of promises Π_X, $A \xrightarrow{-P(T)|\Pi_X} S$, and S provides $+X$ exclusively. For the third case, S is unique and A is dependent on the promise of T. If alternatives existed, only the first case would apply. If A trusts S and accepts its price unconditionally $A \xrightarrow{-P(T)} S$, it might accept an imposition, but it does not have to, because T is not a dependency. Thus an imposition must uniquely supply a critical dependency in both cases.

Promise Theory tells us that imposition is usually ineffective as a strategy for achieving an intended outcome. We shall not speculate on how agents might be coerced into accepting a certain price here, and henceforth assume that prices may always be represented as promises, even it their acceptance is imposed somehow. We note that the promise to accept one price does not preclude the possibility that an agent may accept several different prices from different agents, in other contexts. Having multiple suppliers with acceptable terms is even sound practice for hedging against uncertainty.

We define price initially in full generality, without reference to money:

Definition 73 (Price). *A promise of what a seller agent S will accept in compensation $P(T)$ for keeping its promise to provide a thing T, represented by an acceptance promise to a buyer B:*

$$S \xrightarrow{-P(T)} B \qquad (7.40)$$

This may also be called the 'asking price'. See definition 74.

Note that different agents may set different prices for the same item T. An agent may exact a price in any kind of form, as a penalty or an incentive, with different intentions. Such prices may not be exclusively moneylike, as we shall see, because they attempt to reflect additional semantics, not only amounts[44]. We shall mainly be interested in the kind of price that is most moneylike for a monetary economy; however, to arrive at this notion, we shall need to think carefully about the promises and why a price measured in money makes certain beneficial promises.

However, henceforth we shall focus on prices that are expressed as monetary amounts, as this is dominant. Because money has no semantic labels other than its amount, it cannot adapt its behaviour to local circumstances; it merely transmits a payload (amount) from end to end of its networks. Therefore, the information to adapt to local circumstances must be at the sender and receiver, which implies 'agreement' about the appropriate response to these circumstances has to be encoded in price. This is consistent with the notion of price channel as the primary information channel in an economic system (see also the discussion in section 8.4).

Assumption 11 (Offer price reflects the current context of the seller-buyer channel). *The price offered by a seller is a function of all the local contextual variables on which a seller depends (including perhaps the buyer's offer). Offer price is effectively a sampling of a set of variables local to a seller agent S up to the time of the promise. It may be a Markov process of order n.*

Example 45 (Urgency to buy). *The relative urgency of the buyer and seller to complete a transaction might be viewed as the major influence on price. If agents can do without a thing, they have no need to accept a price. However, the acceptance of a price is also a trust building act, and could be viewed as an investment in future relations. Similarly, if a seller has no need to make a sale, it can wait for an alternative. Clearly time plays a role in price.*

Example 46 (The lifetime of an offer price). *A promise to accept the terms of an exchange has limited validity. Since promises have lifecycles and finite time validity, so must prices. If we deal with the scale of single purchase events, time may play a binary role in the outcome. If we are speaking only of statistical aggregates, then transactional timing places limitations on the timescales over which averages are computed in order to promise stability. Once a transaction has completed, the price at which it completed becomes a matter of history and is invariant.*

Example 47 (Pharmaceutical prices). *The price of medicines (especially insulin) has been a point of contention in recent times. Monopolies by patent rights or effective supply have led to the artificial inflation of prices. Suggestions of influence on government* *regulation by lobbyists have also been asserted. State and private actors negotiate what pharmaceuticals to buy and where, sometimes on behalf of insurance companies, who supposedly act on behalf of their clients. Prices, however, remain largely unrelated to costs of production. High prices are rationalized as providing the incentive for future research in the pharmaceutical industry. However, logically this is only an incentive to invest in research that brings future revenues of equal or greater magnitude.*

Example 48 (Auction bidding prices). *One could argue that the price that results from an auction depends principally on the characteristics of the buyer(s) rather than the seller.*

The terms of exchange are principally, but not exclusively, reflected in price. If a part fails to keep its promises of terms, the other party may attempt to seek compensation for its inconvenience. Some authors may choose to lump together these semantic developments into some kind of an average price, but this presupposes a large statistical scale, so we shall avoid this here.

- Promised delivery time (urgency)

- Window of semantic applicability

- Window of uniqueness (competition)

- Acceptance of loss in lieu of future services.

As always, semantics yield the most important constraints on dynamics, while dynamics underpin what can be supported semantically[Bur13b].

Because a price may be considered a promise, Promise Theory tells us that it is determined autonomously by the selling agent. This means prices are initially determined as *policy* by sellers alone. In any network, in which trust and cooperation play a role, this is not the end of the story, however. The decision to accept the price is made exclusively by the buyer. The effects of cooperation may then impose constraints on a seller, in order to achieve a desired outcome of a sale (or to deter sale, if that is the purpose of pricing). The seller may need to keep other promises, which interfere with the offered price; nonetheless, the ultimate decision of a price level is an autonomous decision by the seller alone. Any suggestion that prices can be settled deterministically violates the autonomy principle, and must be considered false.

7.6 PRICE AS A FORM OF LICENCE

A price plays the role of a licence, or expression of intent: traders should accept money in principle (it provides a licence to buy under social norms), but they may refuse money

of certain amounts, for any reason. When we make an offer of money, we are measuring
the thing being purchased (like weighing a commodity). The amount of money we give
records our assessment or measurement of the promises made by the good or service.

Definition 74 (The asking price). *The price for a product or thing offered to a potential
buyer is a promise to grant the buyer a licence to acquire ownership (the right to purchase
for the amount) X, promised by S. Let a thing $T_a \in \{G_a, S_a, M_a\}$ be a good, service,
or monetary amount in any currency, then price is a mapping of things into an alphabet
of prices:*

$$\mathcal{P} : \mathcal{T} \to \mathcal{M}. \tag{7.41}$$

Definition 75 (The offer price). *The price P_T a buyer promises to pay a seller in return
for thing T.*

$$B \xrightarrow{+P_T|T} S \tag{7.42}$$

We could speak of the bare and dressed offer prices, that include taxes, surcharges, and
levies on transactions by different sources. Even after this amount is decided, further
transactional charges might be levied by the bank of the payer and by the bank of the
payee.

Definition 76 (The final price). *The price P a seller accepts from the buyer for a thing
T.*

$$S \xrightarrow{-P_T} B \tag{7.43}$$

7.7   FINITE INFORMATION PRICING

To describe prices formally, in preparation for defining markets, we take an approach
based on the theory of communication[SW49, CT91], to represent the possible measures
that can communicated as a price in terms of a standard 'codebook' or alphabet.

Definition 77 (Price range alphabet). *Let $\mathcal{P}$ be the set of all possible discrete payments,
and let $P \in \mathcal{P}$ define a partitioning of $\mathcal{P}$ into non-overlapping subsets: $P = \{p_a\}$,
$a = 1, 2, \ldots |P|$. The set p_a defines a digital alphabet for the transfer of information
about price, in the sense of [SW49, CT91].*

Although the values would normally be considered money, they might refer to any *thing*.

Example 49. *Most forms of price would be measured in money, but exchange price could have any semantics, as in the following examples:*

- $p_1 =$ Euro $1, p_2 =$ Euro $10, p_3 =$ Euro $100, \ldots$

- $p_1 = 1 - 5$ USD, $p_2 = 6 - 10$ USD, $p_3 = 11 - 15$ USD, $\ldots$

- $p_1 =$ goat, $p_2 =$ sheep, $p_3 =$ half sheep, $\ldots$

The finite accuracy of the information employed to represent monetary value avoids the transfer of useless or unnecessary distinctions, leaving the alphabet of monetary communications *compressible*. This is an obvious advantage for a network communication technology.

7.8 PRICE AS THE INTENT TO PROBE

An agent may assert a price, based in its own valuation assessments of the thing by either promise or impositional means, based on its assessment of the thing:

$$S \xrightarrow{\ P(T)\ \mid\ v_S(T)\ } B \tag{7.44}$$

$$S \xrightarrow{\ P(T)\ \mid\ v_S(T)\ }\blacksquare\ B \tag{7.45}$$

This price $P(T)$ need not be accepted by a buyer (via a corresponding $B \xrightarrow{\ -P(T)\ } S$). The importance lies in the mapping of the price to a measure:

$$P(T) : T \to \mu(P(T)), \tag{7.46}$$

where μ is what we call a monetary currency. Notice that value does not enter into this expression. It is now redundant. Moreover, since the valuation v_S is made by S, it can be based on any criteria S wishes to employ. The price might reflect an assessment of value under some set of circumstances, but we must also accept that S can ask for as much as it believes someone is willing to give for it, and may also decide to give the item away for nothing, or even pay someone to take it away, due to interfering concerns.

Agents with something to sell may promise an asking price up front, or potential buyers might approach them with an offer first. The order of these promises is not fixed. The promising of a price by the seller indicates an intent to sell, and invites buyers, effectively granting them a licence to purchase at the asking price.

When the buyer counters by promising an amount of money, in trade, this acts as a probe to test the resilience of an asking price, a kind of measuring stick, to measure something about the thing concerned, in the eyes of the seller and other potential buyers. One could try to argue that probing and settling on a price measures the true value of something, but whose assessment of value would that be? If we follow the laws of semantic scaling[Bur15], then the answer is clear: any such price could be interpreted as an assessment of the coarse grained collection of agents involved, reflecting the group rather than any one of them necessarily. However, this has all the usual problems associated with value. In a marketplace of several sellers, competition may distort the mapping between the current owner's perception of value and what price he or she expects to get for it.

It is simpler to bypass these speculations and define the purpose of offering money in trade to be an intent to buy or to sell. This needs some clarification. Why can't we measure the same with a hundred different measures (goats, sheet, wheat, errands performed etc)? Of course, this is possible (it is simply a different alphabet), but it is expensive because every communication needs to be translated for every trade individually. Money offers a logical centralization or calibration of meaning: by using the *lingua franca* of money, we push any price conversions to the edges of an economic network, where every agent can mind its own business, and suffer the cost of its own eccentricity. Thus all agents become homogenized in their promises, and the promises become directly comparable.

The probing of price, in this way, is analogous to pushing on something to see if it will move, until it will move no more. If one pushes a little harder, the price might move some more. Economists adopted the term equilibrium, as used in Newtonian physics, for this balance (meaning literally equal weight). We can follow this nomenclature. When a single buyer interacts with a single seller, we call the outcome of the mutual interaction:

Definition 78 (Type 1 equilibrium). *The mutual information transferred in a promise binding $S \cap R$, in which both parties find a price they both agree to. This is a Nash equilibrium[Nas96].*

When an equilibrium has been reached. Payment may ensue.

7.9 TRANSFER OF OWNERSHIP FOR PAYMENT P

Let's now trivially combine the complementary promises for payment and ownership transfer, assuming that price is predecided. Let the subagent T now represent a thing for sale. Only the directed emission is now relevant. The partial ordering of events, by convention, is now[45]:

1. The good is transferred from A_1 to the new owner A_2 (without precondition).

2. The money is transferred from A_2 to A_1 if the good is accepted.

3. The ownership is changed from A_1 to A_2 if the money is accepted.

In the simplest case that delivery does not pass through any third parties, such as delivery agencies, applying (**Ax3**) the promises take the form:

1. Emission (directed) of an agent T from the body of agent A_1 to target agent A_2 involves the following changes. A_1 deletes the promises in equations (4.33) and (4.34), and replaces them with the following, including an imposition to change the owner of T to A_2.

$$A_1 \xrightarrow{+T} A_2 \quad \text{(deliver good)} \tag{7.47}$$

$$A_1 \xrightarrow{+\mathbf{def}(\text{owner}=A_2)\ |\ P} \blacksquare\ T \quad \text{(impose change)} \tag{7.48}$$

$$T \xrightarrow{+\text{owner}=A_2} * \quad \text{(implemented change)} \tag{7.49}$$

$$A_1 \xrightarrow{-\text{owner}=A_2\ |\ P} T \quad \text{(accept change)} \tag{7.50}$$

$$A_1 \xrightarrow{+\text{owner}=A_2} * \quad \text{(optional)} \tag{7.51}$$

$$A_1 \xrightarrow{-P} A_2 \tag{7.52}$$

2. Absorption by A_2:

$$A_2 \xrightarrow{-T} A_1 \quad \text{(accept good)} \tag{7.53}$$

$$A_2 \xrightarrow{+P|T} A_1 \quad \text{(pay if accepted)} \tag{7.54}$$

$$A_2 \xrightarrow{+\mathbf{def}(\text{owner}=A_2)} \blacksquare\ T \quad \text{(impose change)} \tag{7.55}$$

$$T \xrightarrow{+\text{owner}=A_2} * \quad \text{(implemented change)} \tag{7.56}$$

$$A_2 \xrightarrow{-\text{owner}=A_2} T \quad \text{(accept change)} \tag{7.57}$$

$$A_2 \xrightarrow{+\text{owner}=A_2} * \quad \text{(optionally advertise change)} \tag{7.58}$$

In this version the money P was not owned. We assume, for this example, that ownership of the money remains formally free or with the bank, since it has no validity outside of the bank's boundary.

What is important about this formulation in promises is that the promises are only partially ordered in time sequence. Thus the transaction is not rigidly fragile with respect to sequence. We see the asynchronous nature of payment and transfer that money facilitates. Preconditions are decoupled somewhat, leading to only weak coupling. This aids the time-stability of an economic system by making it more tolerant of latent delays and unforeseen uncertainties. If all transfers requires co-location and simultaneity (spacetime localization) it would place strong restrictions on the scalability and resilience of trade.

7.10   THE INVARIANCE OF MONEY, RELATIVITY, AND THE COVARIANCE OF PAYMENT

Having described both the edges and the body of a network (represented by prices and money respectively), we can now discuss the invariance of money and price more coherently. We have defined money as an invariant with respect to space, time, and the particulars of exchanges. We shall now show why this view makes most sense, compared to one in which money represented an assessment of value.

7.11   DIMENSIONAL ANALYSIS

As an invariant characteristic, we expect a description based only on dimensionless ratios. Dimensional analysis tells us that price P is related to quantity Q or simply N and payment amount μ,

$$Q \equiv N \equiv \frac{\mu}{P}. \tag{7.59}$$

i.e. quantity or number are related to an amount of money divided by price, where price is defined in units of money per quantity of thing. In principle, quantities and numbers are dimensionless here, but in order to account for the semantics of multiple distinguishable things, we can define the dimensions of a thing T_a to be units of $[T_a]$. Then quantity $[Q_a] = [T_a]$ cannot be an invariant, since it can depend on what is being traded. Consequently, neither can price be an invariant. That leaves only money.

We have defined money as an invariant. Since both P, Q are non-invariants with respect to time, promise, exchange, etc, we can now specify more precisely that money is invariant under the following scale transformations.

Lemma 18 (Quantity, price and invariant exchange measure). *Expressing all prices and amounts in money expresses price/quantity relationships in natural units, and money is invariant under a scaling transformation*

$$P \quad \rightarrow \quad \lambda P \tag{7.60}$$
$$Q \quad \rightarrow \quad \frac{1}{\lambda} Q \tag{7.61}$$
$$\mu \quad \rightarrow \quad \mu \tag{7.62}$$

where λ is any scalar constant, or conformal transformation on P, Q.

This follows directly from (7.59). If payment is made in a different currency, we assume there is a conversion matrix (hopefully but not necessarily diagonal), and a scalar Ω that can be applied the edge vectors, in the form

$$P_a \quad \rightarrow \quad \sum_a \Omega_{ab} \, P'_b, \tag{7.63}$$

$$\mu \quad \rightarrow \quad \Omega \, \mu', \tag{7.64}$$

in order to make the conversion. Alternatively, for a fixed sum of money, in any currency, there may be a dimensionless budget transformation Ω_{ab}/Ω.

7.12 NO METRIC SPACE

Let us consider a fictitious 'space', spanned by a basis of vectors $\hat{e}_a$, one for each semantic type of thing that may be bought sold or paid for services. An invariant amount of money μ can now be distributed amongst the possible agents in one of two ways: i) as a stationary potential:

$$\sum_{a=1}^{\dim T} \hat{e}_a \mu_a = \mu, \tag{7.65}$$

with a limited inner product yielding a Kronecker delta:

$$\hat{e}_a \cdot \hat{e}_b = \delta_{ab}; \tag{7.66}$$

or, ii) as a kinetic interaction

$$\sum_{a=1}^{\dim T} Q_a P_a = \mu, \tag{7.67}$$

where Q_a is a quantity of T_a and P_a is the equilibrium price between asking and offer price $P_a = P^{(+)} \cap P^{(-)}$. We might call this invariant interval a budget. This expression has its analogue in the relativity of spacetime physics, where we write $\sum_a dx_a^2 - c^2 \, dt^2 = ds^2$, where the invariant forms the basis for a metric, and a metric space of dimension $\dim T$. Our metric is not Pythagorean, but fulfills the axioms for a metric $d()$:

$$d(Q_1, Q_2) \quad > \quad 0 \tag{7.68}$$

$$d(Q_1, Q_2) \quad = \quad 0, \quad \text{if} \quad Q_1, Q_2 = 0 \tag{7.69}$$

$$d(Q_1, Q_2) \quad = \quad d(Q_2, Q_1) \tag{7.70}$$

$$d(Q_1, Q_3) \quad \leq \quad d(Q_1, Q_2) + d(Q_2, Q_3). \tag{7.71}$$

However, a metric space also needs a faithful mapping into $\mathbb{R}^n$, and we have already mentioned that the finite accuracy of money does not permit this. Thus we can pursue some simple geometric ideas, but must be aware that money does not form a true vector space.

Lemma 19 (Money is not a vector space). *Money, which does not map faithfully (bijectively) to $\mathbb{R}^n$ is not a vector space in the presence of scale factors in $\mathbb{R}$, like percentage fractions.*

In practice, money is a projected embedding into $\mathbb{R}^n$, with varying practices concerning the handling of monetary rounding. From the invariance structure, one is, of course, free to shift the ambiguity from money to prices, which are not invariants. The consequence of this is that prices may ultimately be 'lies', albeit small ones.

In the invariant monetary interval (7.67), price P_a has the analogy of a velocity for things T_a. Since price is a linear amount of money, though not necessarily independent of Q, then by dimensional considerations, we may write

$$P_a \equiv \mu_a^P (Q_a) \, \hat{e}_a. \tag{7.72}$$

It follows that

$$Q_a = \frac{\mu_a}{\mu_a^P(Q_a)} \, \hat{e}_a \equiv \xi_a(Q_a) \, \hat{e}_a, \tag{7.73}$$

so that monetary invariance takes the form:

$$\sum_{a=1}^{\dim T} \mu_a^P(Q_a) \xi_a(Q_a) = \mu, \tag{7.74}$$

where the Q dependence shows a quasi-curvature in the coordinate representation, when one tries to eliminate price as a fixed boundary condition. This is indicative of the complexity that arises when trying to encode too much information into the network carrier (money) itself. The coordinatization in which money measures are constant (analogous to a constant speed of light in Einsteinian relativity) shows that all information enters the network at the edges, which is the standard convention for modelling in mathematics. We could try to identify the price as derivative of a valuation of T_a to a specific buyer B:

$$\mu_a(Q_a) \leftrightarrow v_P \left(T_a \xrightarrow{+T_a, Q_a} B \right), \tag{7.75}$$

but this is not an invariant mapping, as it depends on B and Q_a. Thus there is no deterministic function that maps this into an alternative invariance. The only natural invariant is μ, and representations of floating 'currency values' are only clumsy representations of price variations for exchange.

7.13 PURCHASING POWER OF MONEY

Unlike money itself, then, the purchasing or exchange power of money (regardless of its amount) $Q_a(P_a, \mu)$ is not an invariant measure; it varies even according to the observer and the parties involved, and thus it cannot easily be used as a basis for trusted exchange. Indeed, it might not even be a computable function. It confronts the needs, preferences, and circumstances of an entwined network of independent agents, with woefully incomplete information. Nevertheless, the question of what can be bought for money is the preeminent one to most economists, and seems to lead us back to the mirage of value. We can avoid that once again by referring only to the promises of price, without paying any attention to the individual valuations agents might make. Money allows us access to things whose price can be afforded by the amount of money available to the buyer. This is a question of obstacles and enablers[46]. Agents' many and various arguments of whether price represents a fair valuation of something is a separate discussion whose only outcome is to perhaps alter a price.

Money operates in the context of a network of dependent prices, exchange rates, inflation of debts, and negotiations, all of which may be rising or falling at some rate.

Definition 79 (Purchasing power of money wrt T_a). *An assessment of the expected amount of a familiar and commonplace commodity T_a that can be purchased for a fixed monetary amount. This is measured in the dimensions of commodity units per money unit $[c/m]$.*

The price offered by a seller is a function of the desire of the seller to make a surplus. The remuneration offered by a buyer is a function of its desire to overcome an obstacle, with a surplus. The equilibrium of these tensions can only have a restricted invariant meaning for highly constrained set of circumstances:

- When a single price equilibrium, for a given product T_a, can be applied for all agents across the region, i.e. when the influence of individual buyers is negligible.

- When buyers trust the alternative sellers implicitly.

- When the probability of a rate of sale is sufficiently predictable.

We shall try to give more substance to these criteria in section 8.

Example 50. *The Q_a dependence of the price P_a is expected when buyers still have perceived leverage over sellers. Quantity discounts are common in bulk sales, for instance (enabling a distinction between regular wholesale and sporadic retail prices). Big companies can win drive down the price of smaller businesses because they can always threaten to find a different solution (including buying the supplier). Small businesses enjoy no such privilege, which is why investment is usually needed to prop them up for years until they can attain a minimum survival size. Ultimately businesses are looking for a pension on which they expect to prosper, not to make a rigorously fair algebraic* *relationship between quantity and price that applies to all users. Circumstances vary widely, and all agents act opportunistically.*

Example 51. *Consider a simple case where a bulk buyer is buying a number of software licences (see figure 7.1).*

$$P(Q) = P_1 \frac{(Q_{max} - Q)}{Q_{max}}, \quad Q < Q_{max} \tag{7.76}$$

so the price starts at P_1 for one licence, and goes to zero when $Q = Q_{max}$. After this, further licences are free, allowing a maximum pension from each customer of $\mu = \frac{1}{2} P_1 Q_{max}$. To satisfy (7.76) and (7.59) at the same time, we have two equations in two unknowns, and there is a single solution for the bulk price $P(Q)$, determined by the amount of money spent, $P = \frac{1}{2} - \frac{1}{2}\sqrt{P_1^2 - 4\mu P_1/Q_{max}}$, up to a maximum size of $P(Q) \leq \frac{1}{2}P_1$. This tells us the obvious fact that to get a fixed amount of money with bulk discounts sellers have to resist setting Q_{max} too high. More importantly, this illustrates the idea that terms and conditions (i.e. the semantics) of sales play a large role in what remuneration a seller can expect. Attributing price to 'market forces' is not possible in this case, and any argument to support that would have to be based on averages over market distributions.

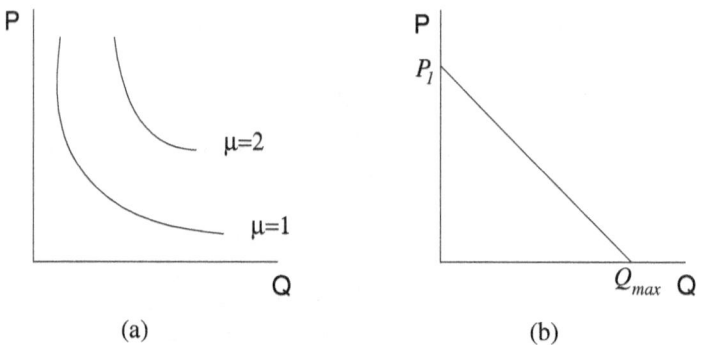

(a)   (b)

Figure 7.1: The general relationship between price and quantity for fixed amounts of money is shown in (a). A specific bulk pricing policy cuts through these curves in (b).

This example is simplistic in its linearity but not uncommon in its essence. Companies are looking for recurring revenues, so there is an incentive to entice buyers back for another round, once the time limit on the good or service has expired. This is a classic game theory scenario [Axe97].

The opposite of a commodity sale is an auction, where there is no premeditated price, instead a duel between several parties to maximize the equilibrium price. Auctions are designed to exploit competition to squeeze buyers. They don't favour customers because they generally start at a minimum price, and rarely go down. Auctions take time to complete, so are assumed to happen quickly in relation to the sales cycle:

$$\Delta t_{\text{sales}} \gg \Delta t_{\text{auction}} \tag{7.77}$$

When an auction or negotiation costs more than the price of a good or service in time or overheads, the thing becomes a commodity and the price is offered 'take it or leave it'.

We need to understand the concept of a market to fully understand the slow dynamics of prices relative to the fast dynamics of transactions. The full story goes beyond the scope of this work, but we lay the groundwork for it. Quantity price relationship is non-trivial for numerous reasons:

- Prices vary between different suppliers for reasons of competition, quality, status, sales campaigns (gambles), dependencies, and more.

- Price depends on costs, including the price of raw materials, transportation, packaging, etc.

- Price depends on negotiation and depends on bartering, quantity, and possibly on variable quality assessments.

- Prices are set to balance aggregate cashflow fluctuations and to build savings buffers to smooth fluctuations in the rate of random arrivals.

- Network effects, such as mutual dependencies, or competitive pricing can create price instability, unless moderated by cutoff policies.

These issues makes prices highly non-linear. For many intents and purposes, prices may be considered to be random variables, but a price sequence over time or some other path variable may not be a Markov process. Memory of past transactions between agents can also play a role in price[47].

Consider a single transaction by a single supplier S of some bulk item T. For an invariant amount of money μ, a buyer B might buy a quantity $Q(T)$. Dimensionally, we can say

$$Q(T) = Q_0 + \frac{\mu}{P(T, S, B, Q, E)}. \tag{7.78}$$

where E represents external information. All we know for certain is that $Q(T) \geq 0$, so that a non-negative price tends to reduce the quantity $Q(T)$, while a negative price will tend to We note that this is not a reversible expression. Changing $\mu \to -\mu$ does not in general allow a sale to be undone. Plotting $(Q(T) - Q_0)P(T, S, B, Q, E) = $ const yields a figure something like figure 7.1.

Example 52. *Complex price semantics arise when promises of goods and services are composed from many parts, each with networked dependencies. A good example of this is how airlines price the seats on flights. They try to predict the future cost based on a variety of promises that may or may not be kept, such a fuel price, demand, etc. It might seem that a seat on an empty flight should be cheap, but costs do not scale continuously. The cost of carrying a single passenger in terms of weight, versus weight of fuel depends on them keeping their promise of baggage allowance, body weight (which they do not promise), and the price of fuel, whose promise changes on a timescale much shorter than ticket sales. This is combined with logistical costs of having planes promised to be at different locations for availability, which in turn depends on weather conditions and a variety of factors that make the pricing a gamble. Like weather prediction, detailed information might enable a brute-force calculation, still with some uncertainty. All these considerations lead to an accumulation or orders that is by no means a Markov process: expectations for the final flight depend on the order and time at which ticket purchases come in. Costs may be unfairly placed on certain passengers at the time of booking, because the semantics of purchase are to promise a price up front, instead of later when the costs are actually known.*

In many cases, as long as there is sufficient stability in the prices promised, fluctuations can be evened out over a timescale much larger than the timescale of ticket purchases and flights. Thus stability is ensured by the thermal reservoir model again, where the size of the reservoir defines a critical scale for being able to equilibrate with impunity. Market monopolies that can aggregate all orders in a single bank buffer will have greater stability.

Scales (timescales) play a central role in the ability to make predictions, even with promises. Moreover, we know that promises will not always be kept, so we must have sufficient bulk redundancy to even out (stabilize) fluctuations, in both dynamics and semantics.

Human responses are emotional in the short run, but may approximate 'rational' when averaged over long timescales and statistical populations (ensembles), by semantic averaging. Thus, if we attempt to model the economy in terms of rational agents, it will lead to management over a timescale much great than that of individual concerns. The economy will not serve us as individuals. Who then will it serve?

7.14 THE PRICE OF MONEY

We have established that borrowing and lending are the principal mechanism for creating money, and that the act of monetary creation has no real overhead cost or risk associated with it for a bank. This means that the space in which money operates ('the economy') may inflate or deflate over time. Nevertheless, in modern capitalism, lending of money is associated with the payment of charges known as 'interest', which can dampen this. This is not a cost, but an imposed price for borrowing money, and demands a description in terms of Promise Theory.

7.15 TIME BASED PROMISES (INTEREST AND RENTS)

As remarked earlier, time plays a particular role in monetary issues. It is the only parameter that does not enter into the invariance relation (7.75) as an explicit agent. Rather it emerges as a characterization of the local state of a network at different scales. It is a relativistic parameter, affecting all promises locally. Let's try to make this more explicit. To make sense of time, we need to refer to clocks that measure it. A clock is an agent with internal states that represent counting[Bur14].

> **Definition 80** (Time based promises). *Any promise to deliver an outcome within a promised interval of time $t_0 < t < t_0 + \Delta t$, as measured by a clock C external to the payer.*

Repeated payments are common, and signal ongoing relationships between agents that build or erode trust. Repeated interaction is the basis of cooperation[Axe97].

Lemma 20 (Time based payments are conditional impositions). *A time based payment is a time based promise to pay. A promise to pay within a time interval $t_0 < t < t_0 + \Delta t$, by an agent A depends on the promise made by a clock agent C, applying (**Ax3**):*

$$A \xrightarrow{\;+pay\;\mid\;t < t_0 + \Delta t\;} R \qquad\qquad (7.79)$$

$$C \xrightarrow{\;+t\;} \blacksquare \quad A, R \qquad\qquad (7.80)$$

$$A, R \xrightarrow{\;-t\;} C \qquad\qquad (7.81)$$

The imposition of an external clock time is accepted by the payer and payee, but they do not choose the time it shows. Thus, both agents (perhaps foolishly) accept the imposition of a timeline for payment. This timeline may not match the timeline by which the agent A acquires the necessary money to pay.

The concept of rent is a price for the right of an agent to use, occupy, or hold something for a pre-agreed interval of time. As noted earlier, ownership in some jurisdictions amounts to little more than rent, as ownership may also be limited by law, and ultimately by decrepitation.

Definition 81 (Rent). *A periodic fee $\mu(T, \Delta t)$ promised in exchange for a licence for an agent borrower B to hold (or otherwise be a tenant of) an asset T, owned by another agent O, for a time interval Δt.*

$$O \xrightarrow{\;+T \;|\; \text{Rent}(T,\mu(T)),\Delta t\;} B \qquad (7.82)$$

$$B \xrightarrow{\;-T\;} O \qquad (7.83)$$

$$B \xrightarrow{\;+\text{Rent}(T,\mu(T)),\Delta t\;} O \qquad (7.84)$$

$$O \xrightarrow{\;-\text{Rent}(T,\mu(T)),\Delta t\;} B \qquad (7.85)$$

Rents may be negative, so that the owner pays a tenant to hold the asset.

---

where $\text{Rent}(\mu(T))$, time interval is a promise body that expresses an amount depending on the asset T, the price $\mu(T)$, and the interval covered by the payment. All these promises are necessary context to call a payment a rent.

Lemma 21 (A rent is a time based payment). *Promises to pay rent or interest are time based promises, which depend on a clock determined by the rent collector.*

This follows from the definition of rent, and the assumption that the rent payer has promised to subordinate its autonomy to accept the timeline for payment by the lender. If the rent payer could decide when the rent clock ticked, there would be no need to pay rent at all.

Definition 82 (Arrears). *A cumulative amount of money equal to the difference between the amount of money an agent promised to pay and the amount it actually paid. Arrears refer to a promise not kept, and the etymology reflects the role of time in the promise, or lateness of payment.*

Some banks require a regular rent from account holders for the pleasure of providing accounts as a service, while others offer this service freely, and charge only for certain transactions. Some banks even pay account holders rent for holding their deposits.

When money is borrowed for an interval of time, one uses the term 'interest' (or the archaic term 'ursary') to mean money paid or earned for the use of money lent. In this meaning, interest is a function that maps one promise (to pay a residual monetary amount of debt) to another promise (to pay or receive a rent). We note that, like rent, interest may also be negative, implying that the borrower is paid by the lender to take the loan.

$$I : \pi_{\text{residue}} \rightarrow \pi_{\text{rent}}. \qquad (7.86)$$

This is characterized by promises rather than impositions, since borrowers explicitly sign the promises to these terms, in the relevant small print, when borrowing. Most countries expect borrowers to understand such terms and take personal responsibility; others do not expect non-technical consumers to necessarily understand what they sign up to and provide protections in general law. The nature of the interest function is a matter for policy. The interest might be computed proportional to:

1. The amount of money originally borrowed $\mu_{borrowed}$.

2. The residual amount $\mu_{residue} = \mu_{borrowed} - \mu_{sum\ paid}$ not yet paid.

3. The residual amount promised but not yet paid on the loan, added to the interest from all current and previous installment promises, not yet kept $\mu_{residue} +$ $\mu_{interest\ debt}$.

In the last case, there is compound interest, as the incomplete payment of a an amount of interest from a payment interval is added to the original debt, regardless of whatever has been promised for the rate of repayment.

This traditional positive form of interest appears as a form of rent, in the first two cases, but it has special status, because the form of payment and lending are the same item (money). The late payment of installments is added to the total debt, instead of remaining a separate semantic issue, leading to *compound interest*. Thus, compound interest leads to a 'double whammy'.

Definition 83 (Compound interest on debt). *The iterative addition of unpaid interest $I(D)$ on residual debt D, where $I(D)$ is computed from the current residual debt, to the residue $D \to D + I(D)$, so that on the next iteration interest will also be paid on prior unpaid interest.*

It follows that:

Lemma 22 (Levy on the inability to settle debt). *Compound interest is a rent (or earning) for an agent's inability to keep its promise to repay debt.*

Framed in this light, compound positive interest seems like a threatening form of extortion by one agent over the other. However, compound interest can also be paid by a bank on deposits, in the opposite manner to reward or discourage the holding of deposits. Clearly, the amount accrued by lending will always be significantly more than the amount paid on deposits, since no agent borrows the amount it already possesses. Also the interest rate on deposits is generally less than the rate for debt.

Example 53 (Compound interest). *Compound interest now presents as the amount of money an agent A must have in order to repay an amount μ of debt to another A' at time* *t. If the interest rate is positive, the amount must be greater $\mu + \Delta\mu$. If the interest rate is negative, it must be less than $\mu - \Delta\mu$. The lender pays the borrower to borrow!*

We can distinguish several semantic cases for basic interest:

> **Definition 84** (Interest on customer loans). *In loans and mortgages, interest refers to a rent (or earning) for leasing money as property of the bank, at a rate I_L.*

Given that the payment of interest (positive or negative) is connected with the delayed repayment of a loan, it may be pointed out that interest has different semantics than repayment.

Lemma 23 (Base interest is rent not repayment). *Interest payments on the original amount $\mu_{borrowed}$ are rents or earnings, not negative debts (repayments).*

Lemma 24 (Compound interest is a mixture of rent and outstanding new debt). *Interest payments on the unpaid residue of the original amount $\mu_{borrowed}$ plus interest on arrears are a mixture of rent and negative debts on arrears. Compound interest = Rent(debt) + Rent(arrears).*

Although interest is associated with the creation of new money (credit) alongside debt, it does not have the status of a pure rent on new money, because interest is also charged on other kinds of debt, such as arrears. The promise of new money in (6.12) was a fixed and immutable event, whereas the amount of money one pays in rent on arrears is unrelated to the amount borrowed, and is quite unpredictable, since the rate of interest and the ability to pay are random variables. Incomplete payment of interest (defined by terms imposed by the lender) leads to new debt.

Lemma 25 (Interest payment does not reduce debt). *The payment of interest leads to no repayment of the loan amount, and therefore no reduction in the next interest payment.*

It follows that:

Lemma 26 (Interest does not reduce the total money supply). *Payment of interest to banks does not reduce the total money supply, whereas payment of debt does.*

Because interest payment does not eliminate debt, the money created by debt is not affected when money is repaid to banks that created it.

Lemma 27 (Interest repayment is an untrustworthy promise). *Unless loans have a fixed interest rate, decided at the outset, borrowers make a potentially unkeepable promise to pay an unspecified amount of interest on loans. Their ability to repay compound interest thus cannot be promised, since the amount is unknown.*

The acceptance of a loan with positive interest, by a borrower, is a gamble against the random interest rate imposed by a lender, and the continuity of possible income. The acceptance of a loan with interest, by a lender, is a gamble on the circumstances of the economic network and on the trustworthiness of the borrower.

Definition 85 (Interest rate). *A fraction I of a base amount of debt, expressed as a percentage, whose numerical value is imposed by the lender L onto a borrower B, as a matter of policy.*

$$L \xrightarrow{\ +I\ } B. \tag{7.87}$$

The duration of this promise is usually unspecified in the terms.

When the borrower fails to repay the borrowed amount μ_{borrowed} in total, interest payments are charged on the extended time for which the remainder is unavailable to the lender. The promise to pay interest in this way is part of the terms of the promise to lend.

- The promise to pay interest is a conditional promise, It is ill-defined, because the rate of interest is not specified. Thus, a cynic might say it has more the status of a ransom or worse.

- The money to quench interest payments cannot be covered by the original loan, since it multiplies the amount borrowed; it must come from somewhere else in the economy, and the payment of interest does not remove any money from the economy, since it goes to the bank as profit not to reduce the debt.

- Debt is amplified by the interest rate, but the original loan is not. Interest adds private debt on top of the promise to repay the loan, unless the rent is paid as it accrues. This is why loans come in different types (e.g. annuity and serial loans) in which one chooses whether to prioritize the payment of interest or reduction of the loan itself.

Deposit accounts may pay interest on money held, subject to restrictions on the availability of the money to account holders:

Definition 86 (Interest on deposits). *When customers deposit money into their accounts from other sources, a bank may pay a bonus or a levy proportional to the currently held amount, at a rate I_D, depending on the sign of the interest rate.*

Inverse rent paid on the deposit to encourage customers to store funds in the bank, or as rent for borrowing their deposits. Deposits are amplified by the interest rate.

> **Definition 87** (Interest on bank borrowing). *When private banks borrow from the central bank to balance their books, the central bank charges a rate of interest, which is the official 'Interest Rate' announced by the central banks. This rate I can be positive or negative.*

These three interest rates are all unrelated a priori, but are linked by the network of interactions in finance, specifically by the need to prevent exploitation of the amplification of money. If one could borrow money and place it as a deposit to receive interest greater than the rate of interest on borrowing, then money would simply grow unbounded without cause. To avoid this $I_D \leq I_L$.

7.16 THE RATIONALE FOR INTEREST

The rationale for interest, in its most basic form, i.e. the rental price for borrowing money has become markedly more complicated in recent years, as the rate has—in some cases—fallen below zero and become negative. This has the extraordinary result of charging depositors for keeping their money, and paying borrowers to borrow—like a charge for babysitting someone else's money.

One rationale for charging positive interest is that lenders are implicitly asked to cover the cost of losing access to the money themselves, for other needs and purposes. In principle, they reduce their own buffer of savings against other economic obstacles or for investing in other opportunities, so that the borrower can temporarily increase theirs. This is a risk for them, and a service provided by the lender. However, if their savings would be stored with a negative *interest on deposits*, it would cost them to keep the money, so there would be an incentive to pass a hot potato onto someone else. Moreover, inflation implies that the future potential of having the same amount of money may be altered relative to new prices, so interest should also compensate for those changes too. But over what timescale?

For banks, the argument for risk compensation is weak, because banks have a licence to create new money for lending with few restrictions, the only caveat being that the amount they are allowed to lend is limited, in principle, to a multiple of the amount of external deposits they hold[48]. Today, with the deregulation of many checks and balances on banking, in key countries, there are ample ways around this simple blockade for wily money manipulators. However, money must also serve consumers, because they are the base on which everything ultimately rests. Today, we focus on the abstract economy over a long timescale, which is not relevant to ordinary peoples' lives (see section 7.18).

The normal assumption is that agents seek a surplus, and must charge for the lending. A different view is that society simply imposes interest as a cultural stratification of society, in the 'interest' of those stakeholders on the receiving end of its wealth[Gra11]. This rationale for the charging of (positive) interest on loans has been criticized, throughout history, as both contentious and even unfair, especially when pushed to unreasonable rates, now referred to as *ursary*[Tur69a]. Those in need of borrowing money are surely those who are least able to afford to pay?[49] Nevertheless, positive interest has an important self-regulating function, albeit a blunt one, encouraging agents to build a buffer of savings, and to discouraging agents from catastrophic unlimited borrowing.

Another semantic role for interest is as an incentive, which seeks to influence and even manipulate the behaviour of people in society, changing their spending and investment habits[50]. Graeber argues that this has been a significant function of debt throughout history[Gra11]. Contemporary fractional reserve banks have been effectively granted an open licence to charge rent on a time-based service by the present day legal systems around the world. Their business opportunity lies in choosing how best to lend, in order to maximize their rent return, subject to a few restrictions on deposits. Where does this return go? It goes to the owners of the banks, which may or may not marginally increase the base of bank deposits for lending to others (a 'trickle down' effect). In effect, interest favours the access to new money by the preferred selection of agents who already have access to large pools or springs of money, and are likely to not miss payments. This means, somewhat paradoxically, that those who have low economic means are discriminated against, and all obstacles may remain in place for them. It's a classic case of preferential attachment in a network[Bar02]. Graphs are unstable to this preferential accumulation of 'wealth', and annealing processes are needed (something analogous to dreaming for memory) that smooth out the inequities and prevent the dominance of singular behaviour[Bur16b], to maintain plasticity.

This preferential lending to those with wealth cannot be understated, as it is a direct example of how money comes with default semantics which can be used for discriminatory behaviour in society. When a region's money promises only a simple amount, the only basis, on which to discriminate in a decision process, is amounts (quantity over quality). Thus, it will always be the case that rich agents are likely to be favoured over poor agents, or perhaps vice versa with altruistic intent, when lending plain money. Money's lack of flexible semantics is a weakness on the ability to constrain economic behaviour.

7.17 NEGATIVE INTEREST

Under recent economic conditions, international money flows have sometimes led banks to try to curb the influx of money—often seeking to isolate the money supply from currency fluctuations or local inflation. To do this, they have imposed negative interest rates. This means everything about interest is suddenly in reverse: the lender is paid to borrow, and the saver is penalized for hoarding—a bizarre twist of fate motivated by inhomogeneous money markets.

Example 54 (Negative interest rate history). *According to Investopedia[Inv17a]:*

- *The Swiss government ran a de facto negative interest rate regime in the early 1970s to counter its currency appreciation due to investors fleeing inflation in other parts of the world.*

- *In 2009 and 2010, Sweden and, in 2012, Denmark used negative interest rates to stem hot money flows into their economies.*

- *In 2014, the European Central Bank (ECB) instituted a negative interest rate that only applied to bank deposits intended to prevent the Eurozone from falling into a deflationary spiral.*

Because lending incentives are focused around banks, negative interest rates (as a policy choice) typically start with bank lending policies, and the shadow world of interbank dealings. This may later impact on consumers if banks choose to extend their policies to regular bank function. As noted earlier, there are many 'interest rates' for different semantic branches of lending activity.

The mechanism for lending between banks is the sale of bonds (a money proxy used in financing), so the bond market plays a major role in economic activity, entirely in parallel to the consumer activity. The two are coupled weakly or strongly by need for banks to maintain reserves for lending, and by the non-linear effects of money supply. The buying of bonds is equivalent to lending money to the bond issuer. The normal situation is that the issuer promises to pay the buyer of bonds (i.e. the lender) periodic interest payments to compensate it for the use of the money; but, with a negative interest rate, the buyer pays more! The interest rate on bond returns means that buyers have to pay rent in order to buy bonds, instead of receive rent on the lending.

Why would any borrower or lender accept such terms? An answer for consumers is that the negative interest rate also applies to their own savings, so it would cost them to hold onto their money too—so there is a partial compensation. Moreover, the interest in one market may be significantly better than the costs of keeping money themselves. There is a risk that agents may simply take their money out of the banking system and stuff it into a mattress somewhere—though this would have limited impact, as there is not enough cash in the world for everyone to do this[51]. In financial institutions, bond buyers may therefore be incentivized to pay someone else to look after their money—hopefully, someone who needs it more, leading to a useful societal investment—thus diffusing the money supply from piggy banks to the open market. This is not as simple as it sounds, however. At the present time, the economic network is in a particularly complex and fluctuating state, driven by political and environmental uncertainties. One might speculate that the simple lever of interest rates may soon be no longer adequate to effect the controls and incentives to hold the economy in a detailed balance.

Example 55 (LETS trading). *The LETS currency trading systems commonly have negative interest rates (the market return of the money decreases). What it amounts to is to reduce the effective role of money as a store of monetary measure, in comparison to its role as a means of exchange. With too much money around, not only the purchase power of money is affected (inflation), but also its ability to function as a store of savings.*

7.18 ECONOMETRIC INTEREST

Although the basic concept of interest is quite simple, the reality of it is highly complex. Because interest is a policy tool, which is inherently time-based, the timescale it refers to is crucially important—and all dynamical phenomena like inflation, employment levels, etc, play a role together in a coherent picture of economic activity. Accounts of the role of interest rates in economic and popular literature are further confused by a varying nomenclature for the multitude of uses for the concept of interest in different contexts.

Any form of interest is clearly a symptom of the dynamical nature of the money network—a way to adjust for the effects of time, by trying to anticipate what one hopes will be a stable and predictable system. The motivation for banks' own interest rates is to stabilize their own money reserves, and maintain what is called 'the value of money', i.e. its potential purchasing power or equivalent monetary measure over time[52], given that prices may rise or fall due to totally unrelated market activities. Governments and central banks want interest rates, as one of the few policy levers they can influence, to stimulate economic activity—encouraging investment to repair and enhance the economic ecosystem, for instance. Banks set their own interest rates, under some encouragement from central bank authority. The market for inter-bank monetary proxies, such as bonds, is thus centre stage in this story, and other aspects of interest are derived from it, in a curious feedback loop which would be unstable if left unregulated. For this reason, interest rate policy is generally a mandate of central banks.

As mentioned above, the interest we may pay on our mortgages is only weakly related to the so-called 'interest rate' we read about in the news. The latter is a kind of average estimate of the state of economy policy, usually referring to a single nation state.

As economists try to measure these phenomena empirically, they maintain measures which are also called 'the interest rate'. We call this the Econometric Interest Rate in order to distinguish it. It is a fictional estimate, which nonetheless has implications for economic policy. The basic or *Nominal Interest Rate* is an estimated rate of interest, taken to be representative of industry practice, with no adjustment for inflation. In other words, it is an approximation to the interest rates that are actually imposed by independent banking agents, based on the assumption that prices and valuations are fixed. However, returning to the subject of timescale, inflation changes prices and valuations at its own unpredictable rate, so simply paying back a loan amount may not be considered a fair repayment—lenders and borrowers may want to be compensated for what they have *effectively* lost or gained, in terms of purchasing power, due to inflation. The *Real Interest Rate* thus measures the growth in actual amount of the loan plus interest, taking an average estimated rate of inflation into account[53]. The repayment of principal plus interest is measured in real terms compared against the buying power of the amount at the time it was borrowed, lent, deposited or invested.

On a short timescale, there cannot be any causal relationship between consumer behaviour (which is a fluctuation to an average) and monetary policy (which is slow, because it is based on statistical averages of expected price changes and wage levels, etc). There is a lag or hysteresis between cause and effect, i.e. between the response of econometric measures and policy levels. Interest is thus based on long timescale aggregates. Banks and governments may attempt to govern the complex network economy by steering average direction, but this can—at best—set an average course in fair weather conditions for institutions and businesses that last long enough to endure those variations in the course. The monetary system today lacks the semantics and the dynamical resolution to be able to stabilize a single individual's life on a human level. Effectively, economics is presently like evolution—concerned with the survival of the species, not the individual. This may be a future aspiration for a better monetary technology to solve in the future.

7.19   SAVING IN ADVANCE VERSUS SAVING IN ARREARS

Without borrowing, agents have to wait to accumulate sufficient surplus by saving in advance rather than in arrears (see section 9.4). This comes with its own costs, which are equally difficult to predict: by the time they have saved enough, prices may have risen and their opportunity may have passed, the fresh produce might have perished, or they might have starved themselves. The ability to cheat time by lending is therefore in the interest of agent in a network that relies on other agents; however, some agents prefer to avoid the encumbrance of a debt relationship to other agents. Cooperative specialization makes networks densely interconnected, and dependency percolates throughout[54].

From the perspective of money management, interest is an incentive for agents to save money and to repay debt. This sounds good from a moral perspective, but it assumes that debt is harmful, when, in fact, the benefits of debt can propagate throughout an entire monetary network, because everyone needs a sufficient supply money moving about to keep their role in the interconnected network functioning. Should any part of the network face an insurmountable obstacle (a strike, a natural disaster, etc), in which its locally dimensionless ratio savings/cost falls below 1, the repercussions could affect everyone else in the network[55]. Moral obligations directed at individuals cannot really address this problem, because they are basically ineffectual (they violate the principle of local autonomy). The presence of a central guarantor of societal continuity (a central bank or government licence) to tolerate or even extinguish debts is the glue that keeps trust alive.

If the purpose of money is to enable and limit access to things in a network of agents, in a fair manner, why don't we simply give money or things away in some fair manner, instead of playing Byzantine games with lending and interest? There are several ways agents can do this.

- Crowd-lending is becoming popular as answer to this question.

- Government welfare is a similar idea, in which the government collects taxes and then uses the means to service a queue of applicants for housing, etc.

This approach to enabling and constraining agent behaviour has been taken in various forms throughout history, such as welfare pensions, rationing in wartime. The concept of universal income for citizens is another proposal for that. A deeper reason for attaching costs to money is to try to create behavioural incentives for buying and selling in marketplaces. There is a general assumption that everyone in society owes money to a bank, and thus reducing interest rates would increase spending, and vice versa.

We make the following conjecture for future study:

Conjecture 3 (Inefficiency of interest as an incentive). *The payment of interest on debt has no basis in fair remuneration for the risks posed by the unavailability of funds to the lender, because in a network one cannot predict where any perceived obstacle might occur, or thence where the money supply is needed. Interest could even skew the inability to pay, by non-local causal dependency.*

To understand this better, we make some remarks about time.

7.20 CLOCKS, THE TIME INSTABILITY OF RENTS, AND MONEY SUPPLY

In a network, all payments are conditional in time. No network of interactions can be understood without time, because each network transaction, which changes the state of the network, is a tick of a clock that changes the state of the network and affects the ability of agents to keep promises[Bur14]. It follows that money cannot be understood without time either. The stability of an economic network thus depends on a competition between independent clocks. Income and rents are competing clocks, pitted against one another in local races, centred on particular agents. These clocks tick independently by the changing states of agents that are connected by promises:

- Rent has a time rate.

- Interest is a penalty for exceeding the promised time interval for payment $\Delta t_{\text{pay}}/\Delta t_{\text{charge}} > 1$.

- The spoiling of goods is another.

- The arrival rate of services is another.

- Competition between others is a network may cause choices and priorities, leading to lost opportunities.

Interest is a function of time over the timescale of repayment installments, which in turn are are shorter intervals than the total duration of repayment. Over this latter interval, compound interest is nonlinear and could theoretically grow faster than the total supply of money available to repay it.

Conjecture 4 (Compound interest is fundamentally unstable). *Interest introduces an explicit instability into economics. It starts a clock that tries to keep money moving. However, its weakness is that it doesn't make the money move in all directions, only to the key hubs (banks), which are therefore preferential attractors for money.*

Although interest does not take money out of circulation, compound interest does lead to a tendency for money to be leeched out of general circulation, by a growing encumbrance of private debt. As time goes on, more and more money can be rendered ineffective by the priority of paying interest. One cannot help but feel that this instability is engineered into the very heart of an interest-based system, causing money to pool at network hubs, starving other less connected regions of the network of fair distribution. The question of whether interest is actually sustainable then follows: it is possible to introduce demands for payment that actually exceed the money supply of a currency at some critical level. The total amount of money (allowed communication), which is available in a network, decides what processes can happen in a certain interval of time.

Example 56. *Consider a small town network economy, in which the local government has releases a total of 10 money units into circulation at the signing of the town treaty. The law says that all payments must be made in this currency, and the governor pockets the amount and opens a bank. Prices are determined by individuals, based on their hopes for the future.*

1. A farmer grows 100 units of potatoes.

2. The governor buys 5 units of potatoes from the farmer, and buys a wagon for the other 5 units from the wagon dealership.

3. The wagon is used to start a transport service, at 1 unit per delivery, to help the farmer deliver his produce.

4. The farmer can pay him up to five trips, but so far, only the wagon dealership has money to buy anything.

5. The wagon dealer pays its two employees 2 units each and the remaining unit is used to buy new wood.

6. The two employees can now buy potatoes from the farmer, but the potatoes went rotten before they could be bought.

7. The governor realizes that people need to get money more quickly, to avoid the problem in the future, and opens a bank.

8. Sally borrows 20 units from the bank at a rate of interest of 1 unit for every 10, every month, and opens a restaurant. Next season, she buys 10 units of potatoes, delivered on 5 trips by the taxi, and waits for customers.

9. Alas, no one else yet has any money at all, so they do not feel comfortable borrowing money, so she is unable to sell to them.

10. The governor has one unit from delivering her potatoes, but her meal costs 2 units, so he can't afford it.

11. Others in the town are working hard to make stuff, but no one can buy it (not because they have nothing to trade, but because they don't have official money to authorize it).

12. Meanwhile, Sally owes money for interest. She can't pay the money she borrowed back, so she is committed to her business, and she hasn't made any money to pay the additional rent on her loan.

13. The economy collapses and people return to swapping possessions directly (after hanging the governor for getting them into this mess).

In this simple example, we see how an insufficient amount of money in the right places prevents a network from functioning. The threat of interest prevents new money from being created by loans, and instead of facilitating transfer, money actually becomes a *throttle on the economy. The townsfolk could simply swap their produce, but by being forced to use money, they are in a stalemate. This is a transport or communication problem, but it is often framed politically to blame those who have little for being lazy or lacking an entrepreneurial spirit, while those who collect rents (like the bank) stand in judgement next to an ineffective lending service.*

We see that, in a closed network, the ability to clear payments within a certain time no longer depends on the local money supply (savings) of an individual agent, at any time. There is a self-consistent network in play (see section 9.5). It is no longer sufficient to trust your neighbour; you have to trust in a complex and unpredictable web of causation. The clearing of payments, within a certain time, becomes increasingly unstable as the supply of money in buffers, at each agent location, falls below some value. Any compound interest in the network must then begin to spiral out of control, because the demands for interest increase, while the inability to pay remains. Ultimately, the total debt could even exceed the total money supply. While this makes no sense, it is not a contingency any currency region believes it has to plan for, thus there is no ceiling on debt.

- If all agents always had sufficient buffer against possible 'shocks', or the arrival of unexpected obstacles, the flow of money could continue without the need for interest.

  The question then is whether such buffers (left unregulated) would allow agents to go off on spending sprees, leading to more debt. Carrying debt is not a problem for society (quite the contrary according to this simple analysis), but charging of interest, and debt discrimination do lead to problems such as unequal access.

- If goods were rationed, there could be no spending sprees. This strategy worked during wartime, and was used by various communist regimes in the subsequent years, but it is considered a punitive violation of quality of life and individual freedoms when applied to common commodities. The question remains whether it could or should be applied to luxuries.

- The proposal of a universal income has been made by many authors, going back to Bertrand Russell [Rus18]. It proposes refilling the buffers of all agents in a uniform manner, by the state, to furnish the best possible insurance against economic collapse. Readers may be curious to know that this approach was the essence of the PageRank algorithm used by Google to calculate the importance ranking of indexed pages on the World Wide Web, to compensate for the 'unfair' accumulation of references to certain pages while others were neglected[PBMW98, BBCEM10].

- Betting on the future availability money, in a pyramid scheme, is the approach used by many corporations and might be called the modus operandi of capitalism. However, there is the builtin assumption that the economy and its money supply can continue to grow forever, which implies more humans and more output, with occasional renormalizations as production costs tend to zero[Rif15, Tof80]. It should be self-evident that this cannot be sustained in a closed system forever.

7.21 MONEY, TIME, AND ENERGY AS ENABLERS

The role of energy to an economic system has been raised by some authors. The energy supply to society is a critical dependence to all activity of an economic nature, so it joins the abstractions of time and money as a prerequisite enabler for monetary processes.

Lemma 28 (Energy supply is a time-dependent critical dependency). *As a critical dependence of all promises, the promise of energy supply E determines a timeline for payment along side time and local money supply μ.*

$$A \xrightarrow{+pay \mid t < t_0 + \Delta t, \ E, \mu} R \tag{7.88}$$

$$C \xrightarrow{+t \mid E} \blacksquare \quad A, R \tag{7.89}$$

$$A, R \xrightarrow{-t} C \tag{7.90}$$

$$A, R, C \xrightarrow{-E} Energy\ supply \tag{7.91}$$

$$Energy\ supply \xrightarrow{+E} A, R, C \tag{7.92}$$

From the perspective of Promise Theory, the semantics of money are just one possible way of making an economy of conditional promises. Energy itself is increasingly being traded as a form of money, alongside carbon taxes and environmental levies. There is little room to explore this issue in this book, but it is worth nothing as another kind of promise that can act as currency.

CHAPTER 8

MARKETS

The concept of a market derives from the existence of agents who are willing to buy a good or service[56]. Information theoretically, one might call a market a *channel* for buying and selling, because it consists of parallel interactions over the same alphabet of prices and offers. Economic channels are discrete channels, and we can designate a single purchase by a channel of bandwidth 1, in units of transactions. Our treatment of markets is in the manner of a superagent making exterior promises. Markets may have many interior functions, which we do not address here, and their exterior promises might be affected by these interior promises; however, without further interior insight, an external buyer cannot know about this relationship and has no basis on which to conclude that a market expresses any kind of consensus, analogous to the role of an equilibrium heat reservoir in thermodynamics. Our treatment does not therefore assume any concept of equilibrium associated with markets, other than the basic statistical stability needed to define promises.

8.1 DEFINITION OF A MARKET

When agents come together to exchange things, they advertise their wares by promising certain attributes and the offer of a price. A market is a channel for wares to be both displayed and selected. Sales are events, which behave as random message arrivals, over these channels. Over time, these might aggregate into patterns and trends. The probabilities and likelihoods of certain events can be calculated, in principle, by observations over space (ensembles) or time (cognitive updates), and by the standard technique of separation of fast and slow variables [Bur13b].

Definition 88 (Market). *A tuple* $M(T_a) = \{S, B, \Pr\left(\pi^{(-)}(T_a)|\pi^{(+)}(T_b)\right)\}$ *which connects buyers and sellers, for the exchange of goods or services. It incorporates:*

- *One or more seller agents* $S_i \in S$ *that promise an asking price (a licence to buy).*

- *One or more buyer agents* $B_j \in B$ *that promise and offer price (intent to buy).*

- *One or more products* $T_a \in \mathcal{T}$, *promised by* S, *which are available to* B *for purchase.*

- *The existence of a non-zero match or suitability transition matrix* $\Pr\left(\pi^{(-)}(T_a) \mid \pi^{(+)}(T_b)\right)$ *whose elements are the conditional probability of buying* T_a *given the offer of* T_b, *where*

$$\pi^{(+)}(T_b) \quad : \quad S \xrightarrow{+T_b} B \tag{8.1}$$

$$\pi^{(-)}(T_a) \quad : \quad B \xrightarrow{-T_a} S \tag{8.2}$$

Rejecting the notion of a fundamental deterministic relationship between average estimates for supply and demand, we do not have to completely discount the possibility of an approximate effective relationship between effective supply and effective demand, at scales much larger than a single sale, and under conditions of sufficient stability. The shift from deterministic language to probabilistic language signifies a shift to embrace the stochastic nature of sales, at the scale of a market, in keeping with modern thinking.

A general purpose market might offer a diversity of goods and services, but it is usual to classify or even partition markets into specialized sub-markets that are aligned with certain kinds of products. In this way we filter out only the buyers and sellers who come together with similar intent. The scaling of agency (see [Bur15]) plays a central role in understanding this, because aggregating individual (micro) intentions inevitably involves approximation, which in turn involves a form of semantic averaging, with loss of information and intent (see the remarks in section 8.5). How similar do products have to be to fall into a similar class? Buyer agents need to assess them as having 'sufficiently similar semantics'. This is subjective, so product categories can only be grouped into statistical classes by defining standards, with with buyers may or may not agree. The definition of a market therefore implies an aggregation policy which simply defines how we choose to group or distinguish items.

Example 57. *A product category 'cola':* $T_a = \{$ *coca cola, Pepsi cola, coke zero, Pepsi max, Walmart's cola, Tab, Dr Pepper, . . .*$\}$. *Apart from semantic labels, it may also make sense to separate products with very different prices into separate categories, and thus separate markets.*

Lemma 29 (Overlapping markets). *From a collection of agents A, we can select $S \in A$ and $B \in A$ however we may choose, so that $S \cap B$ may be non-empty. Thus markets M, M' etc, for any products, may overlap.*

8.2 MARKET SIZE

The definition of market size has a number of definitions in the literature, and is used variously in common parlance. From definition 88 for a market, it has been defined as the number of possible buyers and sellers, i.e. $|B| + |S| = \dim(B) + \dim(S)$ at each given moment. Other authors take a more evidential definition, as the number of realized sales measured over a specific interval of time, implying the number of accepted sale promises. This latter definition has the virtue of being concrete and countable:

Definition 89 (Market size for T in region R over interval Δt). *An assessment of the number of agents in a total region R may be counted by looking at the promises (7.27) in which money is transferred. Let $B_i, S_j \in R$, where $i \neq j$, and define the matrix of outcomes that occur in an interval Δt, relative to an observer agent O.*

$$\pi_{ij}^{buy} \quad = \quad B_i \xrightarrow{\ -right\ to\ purchase\ T\ for\ \mu_i\ } S_j \qquad (8.3)$$

- *Measured in number of sales: The number of sales of T in the defined spacetime region is*

$$N_{ij} \quad = \quad \alpha_{kept}\left(\pi_{ij}^{buy}, \Delta t\right) \qquad (8.4)$$

$$|M_N(T)| \quad = \quad \sum_{i\neq j \in R} N_{ij} \qquad (8.5)$$

- *Measured in money The sum total amount of money accumulated by these over the time interval Δt is:*

$$\mu_{ij} \quad = \quad \alpha_\mu\left(\pi_{ij}^{buy}, \Delta t\right) \qquad (8.6)$$

$$|M_\mu(T)| \quad = \quad \sum_{i\neq j \in R} \mu_{ij} \qquad (8.7)$$

These two measures could, in turn, be used to define a market average sale price:

$$P_M(R, \Delta t) \equiv \frac{|M_\mu(T)|}{|M_N(T)|} \qquad (8.8)$$

8.3 AMOUNT OF MONEY NEEDED IN A MARKET

When new goods are made, how can we know if there a sufficient amount of money in the economy to enable these new things to be bought? Money, previously created, might be locked up, hoarded as savings deposits (potential money), rather than free to transact (kinetic money); or there might simply be insufficient money created to cover the desired amount of economic activity (analogous to there being insufficient network capacity to access data). The problem of too little currency cannot happen in physics, because energy and things are equivalent ($E = mc^2$), but it can happen in economics, because money is a totally independent invention to things. Even the principle of homogeneous accounting is insufficient to create something similar because the equivalent of Einstein's relation would be $\mu = QP$, but price P is not a constant.

Price networks exist, by definition, inside a virtual superagent boundary of a currency region. Referring to figure 3.3, we can try to understand how much money is needed at each moment in time to support economic activity, with the help of a simple thought experiment. A godlike observer could sum up the following:

- The sum of all things multiplied by their price would be an estimate of consumer need at each moment in time.

- Margins for future investments of things that do not already exist.

There are some obstacles with this naive sum:

- There may be no deterministic need or demand for things that currently happen to exist, so the estimate is too large.

- Supply may or may not be correlated with demand over each timescale, but is more likely to be correlated in the long run, so how much of a time buffer do we need to keep?

- Competition for non-existent supply may distort money used, in spite of prices, e.g. by auction. This also impacts the oversupply buffer.

It seems an impossible task indeed to predict how much money society needs to do its bidding. Thus it seems important for banks (or monetary authorities) to be able to create money on demand, by dynamical lending. The question of interest on debt looms over this mechanism though: its side effects may actually render money creation impotent in the worst case. Other approaches are also imaginable: using information technology, analogous to taxation monitoring, one could easily create money without debt and deposit it at key places in a network at different times. This is effectively what happens with awards, grants, stipends, and cash prizes. Universal income for citizens (a basic lifetime pension) has a sound network basis too (see section 9.5).

8.4 MARKETS AS INFORMATION CHANNELS

The information exchanged in product promises, prices, and money allows us to make a very simple definition of a market as a channel for sales. A sales channel is common concept in business, and it scales naturally to any aggregation of agents (buyers, sellers, or goods). In simple terms, any information channel forms from the binding between observed and observer:

$$\text{Observed} \xrightarrow{+\text{source info } I_S} \text{Observer} \qquad (8.9)$$

$$\text{Observer} \xrightarrow{-\text{received info } I_R} \text{Observed,} \qquad (8.10)$$

The (mutual) information which propagates depends essentially on the overlap between what the two agents promise: $I_S \cap I_R$. We can go further an use the concepts from information theory to show that a market satisfies the form definition of a channel[57].

An alphabet Σ is a finite set of symbols. In information theory symbols are usually characters, like ASCII symbols e.g. $\Sigma_{\text{text}} = \{a, b, c, \ldots, 1, 2, 3, \ldots\}$. In our case, symbols will represent fixed promises made by sellers. Any set L of strings over an alphabet Σ may be called a *language*[LP97]. Communication between any two agents (seller and buyer) requires there to be languages with congruent symbols. Put simply, a common language is assisted by having a common alphabet. One may also encode a language using a codebook that replaces symbols one to one with mapped symbols, but we shall not pursue this possibility here, though it might be relevant in future discussions.

Information theory defines measures of the efficiency and integrity of transmission. For the purpose of describing markets, we are interested in how intent is transmitted between buyers and sellers[SW49, CT91].

Definition 90 (Information channel). *A tuple consisting of a source agent S, a receiver agent R, a source alphabet Σ_S, a receiver alphabet Σ_R, and joint probability function $\Pr(Y, X)$, where $Y \in \Sigma_R, X \in \Sigma_S$, measuring the probability of measuring Y at the output, given X at the input.*

Suppose we identify the source S with the collection of seller agents, and the receiver R with buyer agents $R \to B$, in definition 88; next let the alphabet of offers be $\Sigma_S = \{\pi^{(+)}(g_a)\}$, where a runs over all distinguishable offers in S, and similarly the alphabet of choices $\Sigma_B = \{\pi^{(-)}(g_b)\}$. Finally, identify $X \in \Sigma_S, Y \in \Sigma_B$, then we have:

Lemma 30 (A market is an information channel). *The joint probability* $\Pr(Y, X)$, *as measured by the assessment of any observer* O, *of a binding between seller and buyer, is now given by*

$$\Pr\left(Y = \pi^{(-)}(g_b)\,,\, X = \pi^{(+)}(g_a)\right) \equiv \alpha_O \left(\pi^{(-)}(g_b)\,,\, \pi^{(+)}(g_a)\right)/\alpha_0, \quad (8.11)$$

where α_0 *is a normalization, such that* $\sum_{ab} \Pr(a, b) = 1$.

The proof follows by direct association. Concerning the timescales for averaging (which are implicit in a definition of the joint probability matrix), we assume these alphabets to be constant over a stable epoch of the market, so that the rate of change of the market is the rate of change of its combined alphabets.

The homogeneity conditions for transmission of intent were discussed in 2.9-2.11 of [Bur15]. Semantic coarse graining of agents was described in 3.8 of [Bur15]. We define a market by an aggregation policy, e.g.

- Market by price range: Compose the frequency aggregate distribution for things of type T_a over different price bands, mimicking the behaviour of agents in S_i and B_j to aggregate products.

- Market by product class: The frequency aggregate distribution by product type leads to an effective distribution of market supply and demand for things. Here we sum over prices, to get a distribution over different product categories at all prices.

More often than not, economists are interested in specific markets, rather than all possible simultaneous things. In particular, they are interested in how competition between different sellers works and influences prices. Products that make similar promises can be aggregated (semantically averaged) into a category of goods or services. Prices for these similar goods can be averaged (a quantitative average) to yield a 'market price'. These matters are usually hand-waved in economic texts, lacking any suitable descriptive language to add precision to the arguments. Using Promise Theory, it is a straightforward to define markets in terms of the promises sellers make relative to the expectations of buyers.

8.5 No go for Efficient Market Hypothesis

The Efficient Market Hypothesis (EMH[58]) makes the suggestion that market prices somehow contain all the information about context of buyers and sellers [Var15, Kee11]. The following excerpt is from Wikipedia:

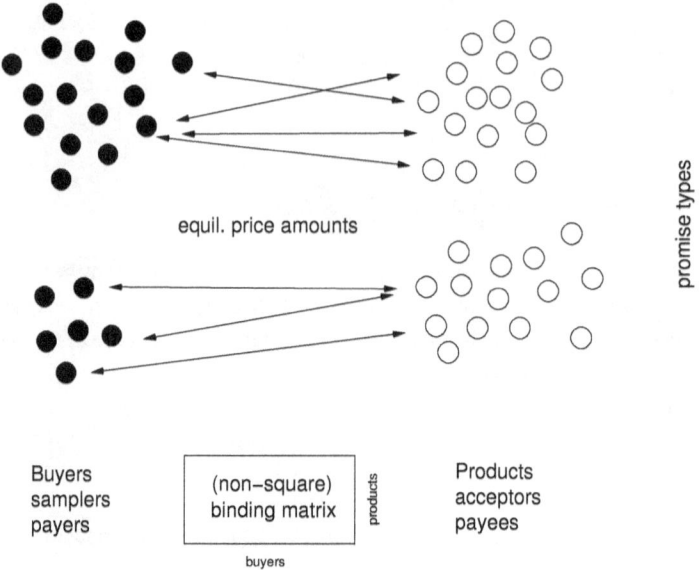

equil. price amounts

| Buyers
samplers
payers | (non–square)
binding matrix | Products
acceptors
payees |

buyers

Figure 8.1: Buyer-seller space versus monetary promise space. Promise space will have clusters of buyer-seller agents and clusters of product (good/service) agents that make similar promises. The graph of the bindings between buyer and product cannot be guaranteed to be a square matrix, except in a world where everyone lives forever, and there are is no product freedom, everything is labelled by the individual who sold traded it, etc. What is more interesting is the arrows between them, and what price equilibrium values they trade for.

The weak form of the EMH claims that prices on traded assets (e.g., stocks, bonds, or property) already reflect all past publicly available information. The semi-strong form of the EMH claims both that prices reflect all publicly available information and that prices instantly change to reflect new public information. The strong form of the EMH additionally claims that prices instantly reflect even hidden "insider" information.

We have not found a more formal expression of the hypothesis than this. What the hypothesis seems to suggest is:

- *All observable contextual variables affecting agents in a market, and their circum-stances, can be mapped congruently into a price alphabet* Σ.

  This seems to be false, because the aggregation of data over space or time must depend on the timescales of the sampling in any information channel, which are not specified. Nyquist's theorem determines a minimum timescale for sampling the complete information, so instantaneous response is impossible. However, since no information can be transmitted instantaneously, we assume that economists intend this to mean 'faster than anything we care about'. Even this cannot be shown, if the rate of change of the information is faster than the sampling process, the information cannot be captured. Thus it would assume that markets change much more slowly than prices, and that prices change much more slowly than trades:

$$\Delta t_{\text{trade}} \gg \Delta t_{\text{price change}} \gg \Delta t_{\text{market context}} \qquad (8.12)$$

This seems to be unlikely.

- *Messages Σ^* in this alphabet can be transmitted with complete integrity and matching message bandwidths to all agents in a market.*

  This assumption cannot be promised, in a network of autonomous agents, as it violates the principle of autonomy (locality) of observations.

Whether we consider from a Promise Theory or an information theory perspective, the indication is that the EMH violates the tenets of information locality.

8.6 AGGREGATE MARKETS, COMPETITION, AND MARKETING SIDE CHANNELS

The question of what we mean by a market depends on the ability to scale intentional behaviour in section 8.9 to much larger numbers of agents (on both sides, buyer and seller). To define markets more carefully we begin with the intent to buy, and scale this to arbitrary size and timescales. Figure 8.2 illustrates how what appears as a simple advertisement of wares, at the level of individual agents translates, into distributions of promise properties and prices at an aggregate scale.

It is how we scale the buying channel that is the most subtle question we have to answer. Let B represent a buyer, S a seller, and g be a good or service promise. The single buyer-seller market was already dealt with in section 7.8. This is a direct channel (peer to peer) from agent to agent. Economists generally assume that markets consist of many agents buying and selling[59]. As we shall see, starting from the assumption of many agents, whose interactions are not clearly described, leads to some problems relating to the loss of information.

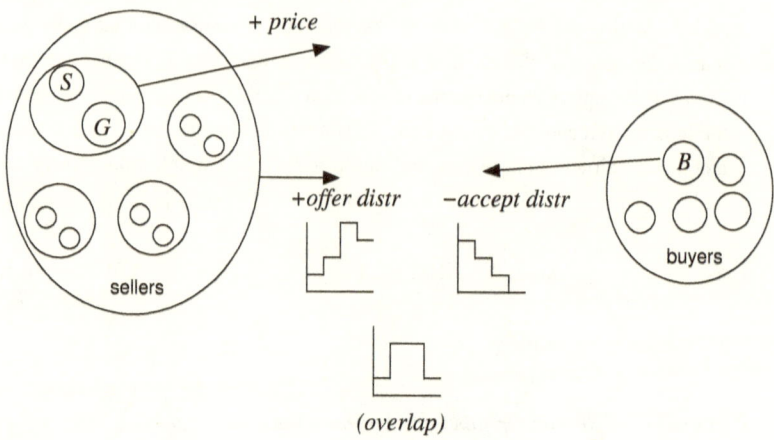

(overlap)

Figure 8.2: A market is an information channel that allows prices and expectations to be communicated about a single kind of product. How do we decide which products are sufficiently similar to belong in the same market? This is subject to semantic uncertainty. Each seller promises its asking price; the superagent of all sellers effectively promises a distribution of all the prices, indicating the probability that a buyer sampling the prices would be offered each price. When this distribution is not singular, this means there is also dynamical uncertainty about price equilibrium, which depends on the scales of time and space we look at. This is particularly true for long tailed distributions.

Definition 91 (Free market for T). *Let B_i and S_i be sets of autonomous agents. All agents are autonomous, and are in scope of one another's promises: We also assume that the promise of product T includes a specification of its qualities and attributes to some degree of fidelity.*

We can scale the aggregate view of a market by replacing arbitrary agents B and S with collections of agents. This can be interpreted as members within B and S at a smaller scale, or more B and S at the same scale. The intent to buy along with an intent to sell then forms a multichannel for possible exchange, initially without further constraint.

$$\{B_i\} \xrightarrow{-T} *, \quad i = 1, 2, \ldots, N_B \qquad (8.13)$$

$$\{S_j\} \xrightarrow{+T} *, \quad j = 1, 2, \ldots, N_S. \qquad (8.14)$$

What is missing from this simple view of redundant channels is *cooperation* and *competition*.

In economic texts, there are many assumptions about competition, 'where a large number of agents compete against one another to satisfy a large number of consumers, and no single agent is supposed to be able to determine how the market operates'. However, little is described concerning the mechanics of competition in an exchange economy[60]. Von Neumann and Morgenstern, Shapely, and later Rapoport have an extensive discussion of competition and cartels in markets, using the coalition approach to games[NM44, Rap66, OR94], which is based on maximization of utility by exploiting the information in the open channels. However, the real question, one might say, is whether the game theoretical models are realistic in their assumptions of agent behaviours[61]. The important outcome, in the context of the present work, is the there is no single market price, but rather a distribution or price vector distributed over the different sellers, whose composition arises from a variety of aggregations. The evolution of a market is the evolution of these distributions[62]. For competition, we only observe the following:

Definition 92 (Competitive T market). *A market for T may be called competitive, when it takes into account promises made between sellers S_i so as to adjust their prices relative to one another, leading to a equilibrium[OR94]. This equilibrium is assumed to happen 'out of band' of the market.*

More important than speculation about competition pressures is the question of whether a market represents a collection of sufficiently *similar* things, or dissimilar things. It is clearly important to a buyer to know that he or she is getting equivalent offers from a specialist market, or even from a commodity market, when comparing prices. How similar or different do things need to be to belong to the same market?

A weakness of price as an information channel expressed in money, is that it cannot represent complex semantics, and therefore has to be accompanies by side channel information ('marketing') that explain the semantics to inform buyers' individual valuations. However, this is also a strength: by separating these concerns from one another, a buyer can easily choose to ignore the different aspects of a 'measure' of a thing.

8.7 PARTITIONING PRODUCT-THING CATEGORIES T_a

In the following, we label types of product by T_a, with index that runs over all possible members of the set T of things. We will often choose to aggregate over these labels too, in order to lump certain products together into approximately equivalent products. Mislabelling may to a distortion of the market information channel, including its resulting price distribution.

In a finite size system, as aggregate agents scale to larger and larger sizes, there can be fewer of them, and the number of interactions they can have is reduced too. Thus in terms of international markets, and vast corporations, one would expect that exterior behaviours to be more like those of individuals in a small community than a faceless consumer in a highly competitive commodity market.

Assumption 12 (Product market partitioning). *An arbitrary clustering of all sellers, within a marketplace, into non-overlapping subsets, based on the category of product promised. Market partitioning is an* approximation *chosen by an observer of the overlap between observed sellers and the observer with an intent to buy (or survey).*

Today, distribution intermediaries decouple products and producers in complex supply chains, leaving only very weak or negligible coupling between originator of a product and its point of sale. If competition involves contention between similar sellers, then competitive markets must become less competitive as the size of agents (companies, currencies, etc) grows.

Definition 93 (Product or commodity market). *An aggregation of agent sellers classified by an approximate definition of a type of good or service promises on offer.*

Exchange or trade is a network property, which starts with the simplest notion of a bond between one seller and one buyer at a time. By the scaling of agency, we can always unify a conglomerate interest as a single superagent entity (role by association).

The relationship between scale and commoditization is important, as we shall explain below. As we aggregate, the special features or individuality of goods and services have to be discarded to cope with the information. Grouping products by type is an approximation, because no two items are truly identical at every level. Thus aggregation is about what information we choose to discard. Thus *information* is the key to what promises enable.

8.8   MARKET PRICE

The settlement on a rationally determined price between autonomous agents is the central problem addressed by Game Theory, and we shall not repeat it, or its connection to Promise Theory, here (see [BB14a]). However, the equilibrium story in Game Theory does not address the question of pre-requisites or the partial ordering of intent, so we provide a brief run through here.

8.9   EMERGENT EXCHANGE PRICE FOR TWO AGENTS (TYPE 1 EQUILIBRIUM)

In order for expectation and offer to overlap, promises offered by both sides must contain some flexibility. In short, there must be an overlap of intent between the seller and the buyer. If both begin with precise deterministic expectations, which are not met, then there must be deadlock (see [BB14a]). In practice, the symmetry of deadlock is broken by a simple protocol, something like the following. For commodities, the flexibility lies in searching for a competing alternative. The seller begins by advertising an amount he or she wants, and the promise of a good for payment (applying (**Ax3**)):

$$\text{Seller} \xrightarrow{-\mu_{\text{want}}} \text{Buyer} \tag{8.15}$$

$$\text{Seller} \xrightarrow{+\text{Good}|\text{accept}(\mu_{\text{pay}})} \text{Buyer} \tag{8.16}$$

The seller might not publicly advertise a willingness to discount some of its 'want' price $\mu_{\text{want}} \rightarrow \mu_{\text{want}} - \Delta_S \mu$,

$$\text{Seller} \xrightarrow{+\mu_{\text{willing}}} \text{Buyer} \tag{8.17}$$

$$\text{Seller} \xrightarrow{-\mu_{\text{pay}}} \text{Buyer} \tag{8.18}$$

$$\tag{8.19}$$

but harbour the intention (promise itself) to accept on policy:

$$\text{Seller} \xrightarrow{+\text{accept}(\mu_{\text{pay}})|\ (\mu_{\text{pay}} \geq \mu_{\text{want}} - \Delta_S \mu)} \text{Seller} \tag{8.20}$$

The buyer, conversely, may initially offer less than the advertised price μ_{willing} but be willing to increase by $\Delta_B \mu$.

$$\text{Buyer} \xrightarrow{-\mu_{\text{want}}} \text{Seller} \tag{8.21}$$

$$\text{Buyer} \xrightarrow{+\mu_{\text{pay}}|(\mu_{\text{want}} \leq \mu_{\text{willing}} + \Delta_B \mu)} \text{Seller} \tag{8.22}$$

$$\text{Buyer} \xrightarrow{-\text{Good}} \text{Seller} \tag{8.23}$$

There is an implicit loop or equilibrium search in (8.20), and (8.22), in which the two parties can meet in the middle somewhere, but not completely deterministically, still with some freedom, iff:

$$\mu_{\text{want}} - \Delta_S \mu \leq \mu_{\text{willing}} + \Delta_B \mu \tag{8.24}$$

This is a simple type 1 equilibrium. The generalization of this kind of interaction is the idea behind rational solutions framed as economic games[NM44, Nas96] (see, for example the discussion in [Ras01]).

Then there is a payment transaction, forming a payment channel:

$$\text{Seller} \xrightarrow{+\mu_{\text{want}}} \text{Buyer} \tag{8.25}$$

$$\text{Seller} \xrightarrow{+\text{Good}|\text{pay}(\mu_{\text{want}})} \text{Buyer} \tag{8.26}$$

$$\text{Buyer} \xrightarrow{-\mu_{\text{want}}} \text{Seller} \tag{8.27}$$

$$\text{Buyer} \xrightarrow{+\text{pay}(\mu_{\text{want}})|\mu_{\text{want}}} \text{Seller} \tag{8.28}$$

$$\text{Buyer} \xrightarrow{-\text{Good}} \text{Seller} \tag{8.29}$$

The negotiation phase of this interaction is usually assumed to happen instantaneously and 'out of band' of the payment channel. However, the negotiation defines a timescale that is non-negligible.

The generalization of this equilibrium is, of course, the Nash or von Neumann minimax equilibrium, used to determine price 'rationally' in terms of a maximization of returns, given a set of arbitrary but fixed strategies (which might be less than rational), and a utility matrix that is also invariant over the course of the game. In a tournament, such as those made famous in [Axe97, Axe84], we also have invariance of strategies and utilities. Thus, without invariance of the domain of alternative choices, rational methods are powerless.

At the end of this section, which dutifully refers to the Game Theory doctrine, we have to ask: what generality can be associate with this assessment of the equilibrium price? Does it apply between any other agents than the two bartering parties? It seems to us that this is of mainly formal interest, as an idealization on which to build a more realistic picture. Of course, the game theoretical problem can be formulated at games of greater numbers of players, but that assumes that all players are engaged in a single competition for each transaction. More likely is the case in which there are many small two-by-two negotiations for price, and the result is a distribution of emergent values.

8.10   EMERGENT EXCHANGE PRICE BY COMPETITION

The theory of games[NM44, Rap66, Ras01] has rational methods to solve these questions, but there seems to be no evidence that such methods are used in the real economy. Rather, price levels are determined by a mixture of expected leverage, cheek, by force of market and powerful organizations, and ultimately by negotiation. We shall not speculate on whether it is possible to explain all the causal mechanisms that result in these prices, and move rather on the necessary outcome, which is a distribution of prices across a market.

Instead, we shall assume that there are two cases: i) either buyers can negotiate with a given seller to change the price, at the microscopic level, or ii) buyers cannot negotiate with the seller, only a choose a different seller, based on sampling the market randomly. We shall argue that the latter case applies large markets, i.e. commodities, where such interactive communication is simply impractical. A stochastic view of price selection makes fewer assumptions than an argument based on rational determinism.

8.11 THE ASSESSMENT OF A MARKET PRICE DISTRIBUTION

A cluster superagent S of sellers cannot efficiently present every variation of price by its interior sellers to buyers, but it can promise information about the distribution of the prices within. Scaling involves such aggregation, i.e. data compression in which a single exterior promise can partially represent all promises on the interior of the agent probabilistically[Bur15]. How could an observer of promise-kept accounting determine such a distribution realistically? We assume the following steps:

1. Sellers and buyers make their promises, defining the players and their intentions to trade.

2. They use or define a finite alphabet of prices P, or ranges (analogous to the ranges in a histogram). Highly detailed prices are costly to use and unrealistic, so there is no loss of generality in limiting the resolution of pricing to certain ranges. We can now count the numbers $\{N(p_a; S)\}$ of promises that fall into each range.

3. Random buyers or impartial observers can sample the prices across the set of sellers, and classify them into a histogram distribution, according to an aggregation policy. There are two distinct policies:

   • Timelike, sequential, cognitive, or Bayesian sampling, taking a single agent sampled on multiple occasions over an interval of time;

   • Spacelike, frequentist, or ensemble sampling, averaging across multiple agents at a single point in time (see figure 8.3).

4. Finally, we define a normalized price distribution by:

Definition 94 (Price distribution). *A price distribution, over a price alphabet P, aggregated across a market M, is defined by*

$$\Phi_{policy}(p_a; S) = \frac{N(p_a; S)}{\sum_a N(p_a; S)}. \tag{8.30}$$

where $N(p_a; S$ represents the number of agents that promises a price in the subrange p_a.

This defines the assumed or estimated probability that a sampling of the market S will result in price p_a, assuming the approximate constancy of the promises. Note that the partitioning p_a is an approximation of arbitrary resolution[63].

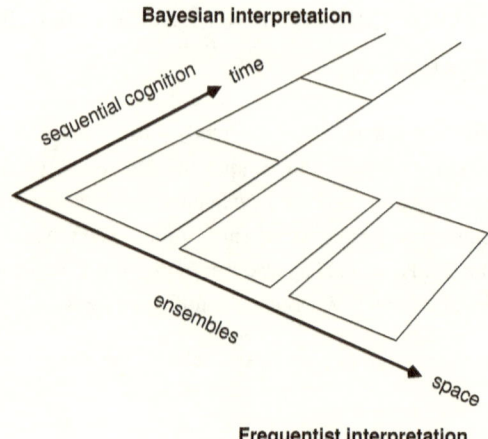

Figure 8.3: A coarse graining policy divides a sample space into assumed equivalences, e.g. different experimental 'trials'. Ensemble statistics represents concurrence (in the same temporal grain), and coincidence (same spatial grain) defines a 'cognitive' or learning experience. In cognition, every new timelike frame is a new experiment, and we must use learning to even out experimental inequivalence.

The price distribution, from the sampling interval, could ultimately by generalized to a function of discrete time $\Phi_{\text{policy}}(p_a; S, t)$, i.e. as a random variable sequence, over a timescale that is much larger than i) the Bayesian sampling interval, ii) the time over which significant changes in price levels can be observed. The behaviour of this sequence may be a Markov process or a memory process, with consequences for dynamical stability. In other words, we learn different things about an interacting system of agents (call it an economy) by observing it across different timescales.

8.12   MARKETS AND ENSEMBLE SAMPLE (A SPACELIKE AGGREGATION)

An ensemble is a concurrent average over many agents all at the same time. It is the principle method in frequentist statistics. We collect all the values of the agents under equivalent conditions, i.e. at the same 'time'. No data are sequentially preferred over others, as in a sequence of barter. There may still be class based weighting of agents.

In such a market, price promises are not usually made one by one to each and every observer, but are made to all agents uniformly in aggregate, i.e. to '*'. The cost of this individual pricing is too high for each agent to determine alone[64]. Thus market prices

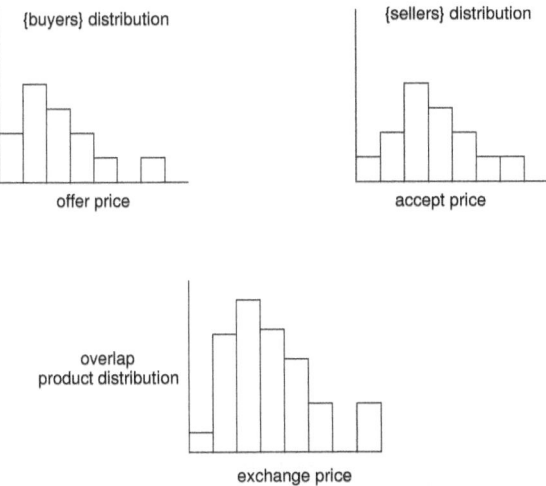

Figure 8.4: Price distributions and acceptance distributions.

may be aggregated as a service, by some observer O, and made available over some schedule of expected stability.

Now, each seller $S_i \in S$ promises its price p_S, and an observer, acting like a price discriminator, samples with and accepts the price if it lies in a range described by p_a:

$$S_i \xrightarrow{\;+p_S\;} * \tag{8.31}$$

$$O \xrightarrow{\;-p_a|(p_S \in p_a)\;} S_i \tag{8.32}$$

where p_a The observer O then assesses whether this acceptance promise was kept and counts the results over all the agents $S_i \in S$:

$$N(p_a; S) = \sum_{i=1}^{|S|} \alpha_O \left((S_i \xrightarrow{\;+p_S\;} *)(O \xrightarrow{\;-p_a|(p_S \in p_a)\;} S_i) \right) \tag{8.33}$$

Now, let $\Phi_{\text{ens}}(p_a \in P, S)$ be a price distribution over a superagent S of agents that promise to sell a single type of product:

$$\Phi_{\text{ensemble}}(p_a \in P, S) \equiv \frac{N(p_a; S)}{\sum_{a \in P} N(p_a; S)} \tag{8.34}$$

8.13 MARKET ADJUSTMENT OR COGNITIVE SAMPLING

A timeseries average is a sequential update over time separated samples, of the same agents. We could imagine this as a form of bartering, or as an adjustment of price due to circumstances over time (no explanation of why the agent makes its promise need be given here). Like Bayesian averaging over successive new inputs. Newer values may be preferred over older ones, or vice versa, determined by a freshness policy.

A particular seller S (which may be a superagent promising a uniform price) promises its price $p_S(t)$, which can fluctuate over time, perhaps due to bartering or environmental costs, and an observer, acting like a price discriminator, samples with and accepts the price if it lies in a range described by p_a:

$$S_i \xrightarrow{\ +p_S(t)\ } \quad * \tag{8.35}$$

$$O \xrightarrow{\ -p_a|(p_S \in p_a)\ } \quad S_i \tag{8.36}$$

where p_a The observer O then assesses whether this acceptance promise was kept and counts the results over all the agents $S_i \in S$:

$$N(p_a; S) = \sum_{t=1}^{|\Delta t_{sample}|} \alpha_O \left((S \xrightarrow{\ +p_S(t)\ } *)(O \xrightarrow{\ -p_a|(p_S \in p_a)\ } S) \right) \tag{8.37}$$

Now, let $\Phi_{cognitive}(p_a \in P, S)$ be a price distribution over a superagent S of agents that promise to sell a single type of product:

$$\Phi_{cognitive}(p_a \in P, S) \equiv \frac{N(p_a; S)}{\sum_{a \in P} N(p_a; S)} \tag{8.38}$$

The assumption here, by Nyquist's theorem, is that the rate of variation in $\Delta t(p_S) \ll \Delta t_{sample}$.

8.14   SUPPLY AND DEMAND DISTRIBUTIONS

By extension of the foregoing definitions, we have:

Definition 95 (Product supply distribution). *A price distribution over a market M, is defined by*

$$Supply_{policy}(\tau_a; S) = \frac{N(\tau_a; S)}{\sum_a N(\tau_a; S)}. \tag{8.39}$$

where $N(\tau_a; S$ represents the number of agents that sells a type of promise τ_a.

Definition 96 (Product demand distribution). *A price distribution over a market M, is defined by*

$$Demand_{policy}(\tau_a; S) = \frac{N(\tau_a; S)}{\sum_a N(\tau_a; S)}. \tag{8.40}$$

where $N(\tau_a; S$ represents the number of agents that buys a type of promise τ_a.

8.15 FIDELITY OF PRICE SAMPLING IN MARKET PRICE

Any coarse graining of agents into roles or categories involves a loss of information (which may be beneficial or inconvenient, in different contexts). We call a collection of agents an ensemble in the statistical sense, because these aggregations lead to statistical characterizations.

The definition of a price distribution, from the aggregation of individual sellers promised prices, makes unavoidable assumptions about who samples these prices and whether this sampling would be the same as one made by a buyer at a different time. The aggregation here involves both a summation over the alternative sellers[65], and a semantic partitioning or classification of the promised prices. The fidelity of this sampling cannot be perfect, because the sampling projects the original promises of price into an alphabet of ranges. This is like measuring sheep in flocks instead of singles.

Assumption 13 (Market approximation). *The purpose of defining a market is to compress the raw information of individual agents and their behaviours into a compressed form that eliminates unwanted detail.*

In other words, the use of market measures may not be the appropriate dynamical characterization of the economic activity, but only helpful facsimile of limited resolution.

The exception to this seems to be the argument that, in the limit of large numbers, i.e. commodity sales, the price adjustment would be small relative to the cost of haggling over price, so the market sampling actually becomes a correct picture.

There is a semantic averaging involved in classifying prices into the subsets p_a, and there remains the question of what price is meant by a range (mean, median, etc). Several of these questions will be answered in section 8.17, in connection with a shift to money.

The assessment of digitizing a sampled price p_S into a symbol range p_a

$$\alpha_O \left((S_i \xrightarrow{+p_S} *)(O \xrightarrow{-p_a|(p_S \in p_a)} S_i) \right). \tag{8.41}$$

Accuracy errors here are weighted according to the relative sizes of the categories p_a. Since the categories are non-overlapping, p_S can only be the member of a single range, so when the sampling promise is kept, $\alpha_O \to 1, 0$, which represents a single bit of information, thus there is a maximum resolution of $|P|$ bits involved in sampling the aggregate price distribution. If the coverage of the sampling is from

Lemma 31 (Market price information lost). *The information lost in a coarse granular price sampling is of the order of the total set capacity $C(P)$*

$$I_{loss} \simeq \log \left(\frac{\sum_{a=1}^{|P|} |p_a|}{|P|} \right) \tag{8.42}$$

where $|P|$ is the number of categories p_a in P, and $|p_a|$ is the number of suppressed members of the subset known to the seller, so that:

$$\sum_{a=1}^{|P|} |p_a| \geq |P| \tag{8.43}$$

As the resolution approaches single bits, $|p_a| \to 1$, $|P| \to \sum_{a=1}^{|P|} |p_a|$, the prices are captured precisely and $I_{\text{loss}} \to 0$.

8.16   INVARIANCE OF COMMUNICATED INTENT

Both the individual agents, and the aggregations of sellers and receivers can perform this kind of assessment. Any local fluctuations in circumstances can therefore to a fluctuating distribution of what sellers are intend to charge, and what buyers intend to pay. The probability of a sale at a price within p_a is the overlap of these distributions (see figure 8.4):

$$\Pr_{\text{market price}} (p_a) = \Phi_{\text{ensemble}}(p_a; B)\Phi_{\text{ensemble}}(p_a; S). \tag{8.44}$$

Under a change of price units, this would become

$$\Pr_{\text{market price}} (p_a) = L_-(\Phi_{\text{ensemble}}(p_a; B))L_+(\Phi_{\text{ensemble}}(p_a; S)), \tag{8.45}$$

which means that invariance (or integrity) of communication in the market requires, schematically[66]:

$$L_- \cdot L_+ = 1 \tag{8.46}$$

This coordination means that L_- is the inverse of L_+, thus they have to be coordinated by a common parameterization $L(\mu)$. This further means that, up to a local transformation individual to every agent, invariance of the intended communication implies a common alphabet for all agents. We can identify that alphabet with money, whose consistency is enabled by centralized handling of a trusted intermediary (i.e. a bank), or by the memory stored in 'social convention'.

This is related to the problem of games of zero (or constant) sum, where it is shown in 15.2.5 of [BB14a] that agents must adopt a common currency in order to cooperate in agreeing on or equilibrating their valuations. Thus, from the construction proven there, we can infer the following equivalence:

Do we really care about the invariance of this probability? Certainly the world would not suffer a major blow if small errors crept into large markets. However, we must remember that we have artificially separated out a single product category for the argument, in pristine isolation. The influence of errors on the total economic situation could be large:

- If we apply this to a market with a small number of large agents, then changes would be amplified.

- The network effect of markets that depend on other markets has not been considered at all, and may well lead to non-linear effects, again with large amplification of the result.

The suggestion of these considerations is that the need for invariance of price communication across markets for stability across the whole network of goods and services, thus practically leads us to invent money as a necessary condition for conservation of probability in market price, and thence a key prerequisite for conservation of money.

8.17 MONEY AS AN EXCHANGE LANGUAGE

There are two reasons why money is practically essential from a network perspective.

1. Agents interact with other agents in a peer-to-peer fashion in a multitude of ways, based on their needs and capabilites (demand and supply). They form a semantic network of rich diversity. However, from the perspective of 'reach', the semantic specificity of individual bindings is a hindrance rather than a bridge between agents. In order for influence to 'percolate' through a network[BCE04, CEM07], i.e. to form paths that span the entire diameter of a network, it is well known that local semantics partition graphs in a way that makes this practically impossible[Bur09, Bur12b, Bur16b, Bur17b]. Unless communications between agents are ubiquitous and without prerequisite types, there will be only highly limited range of interactions. Economically, this means that agents who trade specific goods, without a universal interchange language (money) will have very limited possibilities to support trade, and must have each others' needs fully covered. This represents a network of generalists, not of specialized industries.

2. The fidelity of communicated exchange prices, challenged in the previous sections, lead us to conclude that distributed assessments, made by autonomous agents, may be an unreliable method of communication, unless there is calibration of the alphabets at the end points. Alphabets do not necessarily have to match one to one in order to transmit information with integrity (bijectively), but they do need to preserve the congruence of the association[67]. Relative price relationships at the sellers need to be reflected by the same relative prices seen by the buyers. This means they need to be related by at worst a linear transformation.

If agents are able to use a trusted lingua franca (money), then communication is based on the information channels formed between congruent prices for a random network of demand and supply. If one seller alters is price alphabet by transforming $\Sigma \rightarrow T(\Sigma)$, the other must match this change, else information will not propagate without error. Without a common interloper, agents would be limited to direct pairwise exchanges.

By insisting on the use of a calibrated alphabet, agents engage in a 'network bus' or core trunk exchange network, much like a power grid. If A_1 needs something from A_2, but agent A_2 has nothing A_1 wants, A_2 can receive interchangeable money and get fulfill its needs elsewhere, as long as the balance of payments evens out in the long run. By 'long run', we mean before any agent runs out of money, because if any agent runs out of money, it detaches from the network and can never rejoin without being 'bailed out' as an act of charity or conquest.

Both these mechanisms refer to the isolation of agents from a cooperative network by failing to establish a communications channel to arbitrary buyers or suppliers. In terms of the linguistic constraint on money, as a network transport mechanism, we require some basic properties:

- Any redefinition or scaling of the price alphabet used by the seller must be matched by the buyer. If the new units have different resolution, this might affect the choice of the buyer to accept a price, but it will not affect the transmission criteria for the information will be invariant[Kul68]. This should be intuitively clear. A change of units might alter the values, but if both parties change in tandem, the intent transferred must be preserved by the channel. If the new pricing scheme cannot represent the old price with perfect fidelity, then the seller has to select a new price that it can represent. Similarly, the buyer has to accept limitations in the precision of any discounts it might negotiate.

  Example 58. *If one party chooses to change its price from being measured in sheep to goats, then so should the other party. The new prices may not be the same, but one will be a linear transformation of the other, up to rounding errors.*

- The price alphabet issue relates to transmission, and does not preclude either party from representing its offers internally in whatever units it sees fit. However, by performing all conversions into a common language, at the edge of the network, any agent can avoid trivial barriers to accessing parts of the network caused by an unnecessary inability to communicate its promises.

- Autonomous agents' interior states cannot be coordinated without an intermediate common agency, so assessments of internal states and valuations cannot be communicated without error; however, agents can work around this limitation by simply promising an invariant price, as a facsimile of whatever value they believe in. Like an assessment, this price need not be justified; it has the status of an observation of policy whose selection criteria is unknowable. The success of the choice of price, as an intended strategy, is a totally separate issue about which we can say nothing up front.

It is the dispassionate invariance of price, measured in money, that decouples trade from complex internal semantics of agents and removes barriers to exchange. If other attributes are included in a trade, it decreases the moneyness of the offer, by essentially imbuing the money proxy with additional promises. We can show this as follows.

Finally, if we consider the overlap of alphabets on each end of a price channel, we see why bartering in semantically distinct goods is not invariant. A promise binding matrix from buyers to products is not a square matrix:

$$
\left.
\begin{array}{l}
\text{Product}_a \xrightarrow{+\text{attributes}} \text{Buyer}_i \\
\text{Buyer}_i \xrightarrow{-\text{attributes}} \text{Product} \\
\text{Buyer}_i \xrightarrow{+\text{exchange value}} \text{Product} \\
\text{Product}_a \xrightarrow{-\text{exchange value}} \text{Buyer}_i
\end{array}
\right\}
= \Pi_{AC}, \quad
\begin{array}{l}
a \in \{\text{products}\} \\
i \in \{\text{consumers}\}
\end{array}
\tag{8.47}
$$

Thus, the promise matrix is not suitable for matching offers and acceptances because the domain of its distribution is not invariant over the spacetime span of a market. The domain and range of these categories is simply not homogeneous enough to span an entire ensemble of agents on either side (buyers or products).

To describe an equivalent ensemble of circumstances that unifies a collection of agents into an invariant market (superagent) price distribution, we need to have the stability of invariance over the domain of the distribution function. The domains of asking price and acceptance promises thus need to be invariant across any comparable transactions, averaged over many different agents that we are collecting together as equivalent under some market criterion. Only then will we have stability of accounting. All of this calls for an singular invariant alphabet, which is satisfied by money.

To see this, one could simply postulate the existence of an alphabet of invariant isomorphic states, like a menu of choices agents can use to explain the magnitude of their exchanges. In terms of which both can communicate their mutual offers in a one to one mapping. We postulate the existence of such an alphabet:

Conjecture 5 (We need the invention of money to scale markets with stable price). *Let $P = \{\mu_1, \mu_2, \mu_3, \ldots \mu_p\}$, be a finite set of amounts, measured in units of money μ, where p is the dimension of the domain, and P is invariant under exchanges of equivalent agents, and transactions on the timescale of a stable market. Then any alphabet P is necessarily equivalent to money (as we have defined it) in these denominations.*

It is interesting that it is not money, which is the language of the interloper in trade, but rather price levels. However, both are quantities that can only be compared by use of a common system of units μ, which is a key function of money.

The picture that emerges from a network view of the economy is not one of agents being directed (deterministically) by an invisible hand that ushers in stability and optimality for all agents, as in the pre 20th century view, but rather as a fully modern stochastic network process, calibrated by a global patchwork of regional meanings. One does not escape the role of semantics in such an interpretation. Agents effectively sample random variables and form patterns across the aggregate scales of semantic similarity. This is the modern information theoretic view of economic exchange, which is certainly analogous to other statistical and non-deterministic modern descriptions of the world, including quantum mechanics and statistical mechanics.

8.18   THE SIGNIFICANCE OF AGGREGATE SCALE

The efficiency of any agent's influence over buyers' and sellers' intentions depends on the sizes of the markets they are engaged in. At some scale it becomes a losing strategy to try to negotiate on the price of certain individual things at the retail level. These are the goods and services we call *commodities*. It is true that wholesale distributors may engage in auctions and bartering of bulk purchases, by treating an aggregation of promises as a single bundle, and thereby wiping out any distinction between individual variant representatives. Such semantic averaging is the penalty in lost information (and perhaps) money of treating similar things as indistinguishable.

A self-stabilizing set of conditions may arise, when goods and services have a low specificity value to buyers. Aggregate groupings converge like a *de facto* standard and promise 'mass market' appeal. This makes them both easier and cheaper to mass produce, and to easier to sell by matching to blunt requirements. Eventually, the optimization of such commodities for market channels must lead to a self-consistent 'price race to the bottom', as follows:

- When a product's promises are accepted by a large number of buyers $N_B \to \infty$, the promises it makes cannot be specific to each buyer, as the cost of customization would grow in proportion to N_B, so the information content promised tends to a constant, which makes goods easy to copy.

- When the number of sellers of similar products is large, there is a simplification for buyers too. The indistinguishability of products means that the value offered by all sellers is equal.

- The cost of negotiating a price reduction would be $\sim$ constant$/N_S$ for the buyer, which tends to zero as the number of sellers becomes large. For the seller the cost of negotiation grows with the number of customers constant $\times$ N_B, so sellers are motivated to avoid negotiation by offering a low price. As the number of sellers grows, competition thus favours a sharp distribution $P(p_a)$ around a single price $p_a = \epsilon$, so that the price distribution is a delta function $P(\epsilon) = 1$. The indistinguishability of sellers' promises reduces the cost of finding the best buyers (any seller will do).

- If the price of a product $P = \epsilon$ can be small relative to the cost of negotiating or trying other sellers, this offers the incentive to not negotiate a lower price, i.e. the variation in price across different sellers is negligible so nothing is gained from going elsewhere, the cost of sale is low which allows the price ϵ to be low, and efficiencies of scale allow further savings on the seller side. As the number of sellers grows, the price must get even smaller to satisfy this, so as market bandwidth

$$N_S \cap N_B$$

  grows, the price ϵ must shrink to a level at which buyers don't care about the differences.

- This is consistent with the cost of sale for the seller being reduced as N_S grows, because the likelihood of bartering will scale approximately inversely like

$$\text{constant}/N_S$$

  . The profit on a sale might be low, because ϵ is small, but this is compensated by a large N_B, and efficiency in production.

Notice that this restores the type 1 equilibrium price, where a buyer simply accepts an offered price or doesn't, because the cost of negotiating rounds for type 2 or 3 is much higher than the perceived value of the product. This effect of a competitive equilibrium puts pressure on sellers to lower their prices, when the products are simple enough to mass produce for a general buyer market.

This self-consistent set of conditions is what we shall define here as a commodity.

Definition 97 (Commodity). *A good or service, assessed by buyers to have a low value, and for which the expected revenue tends to zero, even as the cost of production tends to zero itself, because the good is of low information content, making competitor versions indistinguishable, leaving only price as a distinction.*

$$\frac{Cost\ of\ sale}{Return/exchange} \to 0 \tag{8.48}$$

as

$$Cost\ of\ sale \to 0 \tag{8.49}$$

$$Return/exchange \to 0 \tag{8.50}$$

This assumes a very large market bandwidth.

Example 59. *Attempts to rebrand water by bottling and adding gas, flavours, etc, adds semantic labels to the commodity in the hope of making it non-interchangeable so that a new price can be negotiated.*

Our definition matches quite closely the conditions for so-called *perfect competition*, as described in most economics texts. In fact, it seems to us that economists assume that all products are commodities and behave in this way. Our purpose here is to distinguish this kind of process from niche specialist products, which necessarily have to cost more and are based on a cognitive learning rather than an ensemble averaging process.

Commoditization of goods and sales leads to the elimination of human parties with impartial exchange. This is analogous to the replacement of peer to peer interactions with centralized (dehumanized) services, as discussed in [Bur17a]. In this case, prices are generally fixed, or agreed by automated auctions. An ensemble market cannot learn cognitively over time as it is expensive to combine timelike development for a spacelike ensemble. Moreover, the timelike changes (Bayesian) averaged over time would likely get washed out by the ensemble variation, so it would likely be self-defeating. Commodities thus try semantic averaging over a broad ensemble in the hope of binding to as many as possible with a 'lowest common denominator' or minimum viable product approach (see figure 8.5). Specialized niche products try to retain a few high value bindings by appealing the particular benefits. The investment cost is high, so these need to give a high return to sustain the market.

A final technical note: can money be a commodity? Authors on the subject of economics often claim it is just another commodity, but here we are interested only in technically defensible statements. It should be clear, from the foregoing discussions, that money itself is not ownable, and can neither be bought or sold (any more than a key can be opened by a key). However, any proxies of money, which promise additional 'value

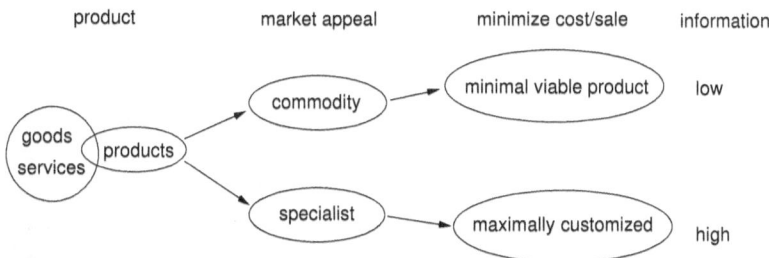

Figure 8.5: Commodities are ensemble averages, designed to be the semantic average that can appeal equally to all. They are mass-market strategies, and the pressure is on low price: low investment and low return. Niche goods invest in cognitive learning: forming a long lasting relationship with long term customer - high investment and high return. As Toffler pointed out[Tof70], the cost of customization has fallen drastically, and commodities are increasingly automated.

added' semantics, can be packaged, assigned prices, and sold freely. These might be either specialist products or commodities, depending on the universality of their appeal. Mortgages, for example, might naturally be called a commodity, as might a bulk purchase of Euros relative to a foreign currency. However, these examples are not just 'money', in context, they only contain money.

8.19 MARKET ADAPTATION BY COGNITIVE LEARNING

Markets can learn from changing circumstances by reflecting changes to market context into non-Markov price variables. Learning is a process by which probabilistic estimates are updated with new information on regular sampling. This requires a system with memory (i.e. not a Markov process). The adaptation of markets can also be viewed as a form of cognitive (Bayesian) learning[Bur16b]:

1. Price adaptation: $\Phi_a(t) \rightarrow \Phi_a(t') = \sum_b L_{ab}(\Phi_b(t), N_b(t'))$, i.e. can mix categories as time evolves.

2. Marketing, branding, design, adaptation:

3. Attribute/promise adaptation (type)

The efficient market hypothesis suggests that price alone can represent all information (see section 8.5).

8.20   OTHER 'MARKET' COMMUNICATION CHANNELS

Prices are the communications channel purported to represent commodities, but these are not the only channels. It seems obvious to us that rich marketing information, including logical argumentation, associations to buyers' backgrounds, and so on, cannot be faithfully represented as price alone, because price cannot make semantic distinctions. It seems therefore that there is no question that selling involves additional sidebands of information, along side price, which must play exactly the role that each individual buyer agent chooses to heed from them. This follows from the principle of autonomy. For brevity, we shall not discuss this further here.

8.21   AGENT BEHAVIOURS IN A COLLECTIVE MARKET

In Promise Theory we consider cooperation to be a generic designation by which agents interact, but it is more normal to view cooperative as a kind of beneficial teamwork:

Definition 98 (Cooperation). *The selection of promises that voluntarily and mutually favour the outcome of a collection of agents.*

Definition 99 (Competition). *The selection of promises by an agent A that voluntarily and mutually favour the outcome of A (i.e. 'self'), potentially at the expense of non-self.*

Definition 100 (Altruism). *The selection of promises by an agent A that voluntarily and mutually favour the outcome of $\neg A$ (i.e. 'non-self'), potentially at the expense of self.*

Altruism may indirectly benefit, as is argued by evolutionary biologists. So-called reciprocal altruism may ultimately favour in the long term[Axe97, Axe84, Bur13b].

Definition 101 (Monopoly). *A state in which a single agent dominates the sale of a particular thing.*

Large size favours economies of scale (sublinear scaling of costs), and may even result in additional benefits, (superlinear scaling of output), as shown by [BLH$^+$07, Bet13, Bur16a].

Example 60. *The evidence for mixed urban areas is that additional size leads to additional innovation, but for partitioned (silo) organizations in a single market, the effect is potentially the opposite. State owned (nationalized) services and industries may lost* *the ability to offer their services to parties outside of the local market they are entrusted to serve. Thus the advantages of economies of scale may be offset by an inability to innovate or raise their market share by selling services to a larger market. In the past, national markets were expected to stand alone, but in the global economy, an industry limited to a single nation might be throttled by this limitation.*

CHAPTER 9

REASONING ABOUT MONEY

We have expended some effort to make some clear and precise statements about money, relative to buying and selling. How can we apply this? What kinds of stories can we tell about money? What questions could we now formulate and try to answer? This topic is too large to fit into this margin, but a few topics seem to call out for attention, including the following:

- What is the causal influence of money in society?

- Can we quantity timescales precisely and encode these into the semantics of money without ambiguity?

- Can we fully describe the meaning of relativity and 'invariance' with respect to money?

- What is the role of a central bank have in determining how much money there is, and what it is worth?

- What is the relationship between money, inflation, and employment?

- What is the role of interest? Why do we really pay it? So we need it?

- What role might semantically specialized currencies have in the future? The way we pay today has direct consequences for particular parties:

    - If you pay with Mastercard, Visa, etc, we'll have to charge you 5% handling fee.

    - If you use ApplePay, there is a 30% commission.

- If you pay with Alipay or TenCent, you might get a bonus.

- If you use air miles to rent a car, insurance is not included.

- Cryptocurrency is risky. If you lose the keys, you lose the money!

There are plenty of examples, any of which might be in flux at any moment.

- Is it possible to guarantee or at least promise the stability of a network of exchange measures?

- What effect will the next generation of high speed reliable communications technologies have on the ability to clear payments faster?

- Is faster and faster clearing actually desirable? If non-linear money works faster than the human mind, can it get out of control?

Several of these issues go well beyond the scope of this book, but we make passing remarks on a few points of obvious interest in this final section.

9.1 THE IMPORTANCE OF TIME

We have observed throughout this work that time is inseparable from money. Money, price, and payment, are interwoven through the promises and conditions for clearing or payments, and the accessibility of money. As the scale of small payments increases by population and the explosion of automated smart services, the burdens on payment systems will grow. Today the general public operates on assumptions about payment, which have not changed much since the 19th century, but the speed of communications has altered significantly. Payments could be cleared faster, but might we lose the causal connection between human intent and monetary behaviour? Clearly this happened at some level during the financial crisis of 2008 to some extent.

Example 61 (Time horizons on repayment). *Interest rates and time limits on access to money have a manipulative effect, but perhaps not the one intended. A time limit may be implicit (as debt becomes unmanageable) or explicit (a promise as part of an agreement).*

Why is there a time limit on loans, e.g. by contract or by accumulation of positive interest? Who is protected by a time limit on a loan with negative interest? If a lender (say bank, and by extension government) says it doesn't want to support access to funds (insisting on surplus within a short time horizon), what it is really saying is that it wants to constrain the freedom on the sanctioned activities of the workforce. By putting all the risk of failure to repay onto to borrower (for whom it matters a lot), instead of government (for whom it doesn't matter much at all), a government declares mistrust in entrepreneurs. There is a disincentive for starting new businesses, unless they can be immediately profitable. This means the workforce has to find different jobs already offered by others. This could be intentional to make people return to 'standard' jobs and stifle innovation. Or, if those concerned cannot promise the right skills, it will simply make them unemployed (and perhaps unemployable).

By lending and increasing the time limits on borrowing, obstacles can eventually be overcome by innovation. If not, this access to money keeps people in employment, where they spend and make other retail businesses successful. It is a simple mechanism to redistribute wealth.

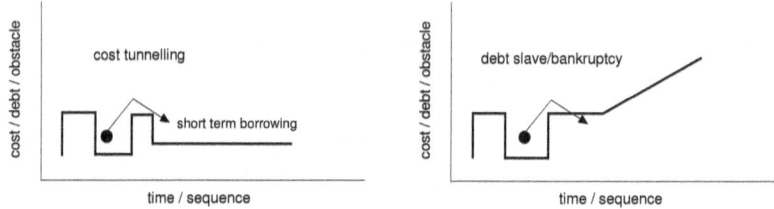

Figure 9.1: The effect of relative timescales on the mechanics of financing. If access to money (e.g. short term loan) has a time limit that exceeds the obstacle, debt can be repaid quickly. However, if compound costs from interest are too high, the obstacle may never be overcome. The situation is analogous to tunnelling in quantum mechanics.

9.2   PURCHASE VS SAVING (RIFFLES AND POOLS)

The dynamical stability of a network in motion is a difficult dynamical problem indeed. In the distant subject of physical geography, there is an empirical fact about rivers, which observes rivers have alternating sections of fast moving rapids (or shallow riffles) and slow moving deep pools. The distance between these is observed to be 5-7 times the width of the river at each point. This is based on simple dimensional arguments within a highly non-linear flow problem. A physicist would naturally try to look for such effects in a monetary flow too. Of course, a monetary network is a much harder problem, and the circularity of flows adds additional constraints that are much more exacting than a mere river flow. But there is an important similarity: when we include time into a description, we must accept that different relative rates of activity between agents will lead to fast and slow points in a network. In computer science, this situation is modelled as networks of queues.

In economics, the analogous topics include:

- Savings and investment (pools or buffers of money)

- Velocity of money (riffles or flows)

- Monetary stability of an economy (will money disappear into a sink, or overflow its banks?)

By appealing to the view of money as a network of promises, we can offer some limited answers to these questions immediately, from the study of graphs[BBCEM10].

9.3 BALANCE OF PAYMENTS IN A NETWORK

In electricity, it is Kirchoff's Laws that apply the conservation of current to what happens at a network junction. If we accept the accounting (trust) principle of intended conservation of money, then Kirchoff's laws must also apply to money. Very simply, they say: what goes in must come out (or remain inside until later). An economic system, on the other hand, is an ecosystem, i.e. a network of interacting agents. Interactions lead to events, which are changes that may be counted as ticks of a clock. Such processes define timescales, and the stabilization of any system, whether mechanical, statistical, or economic, depends on there being sufficient time for interactions to count towards measurable effects.

Timescale	Notation
Single interaction or trade fluctuation	$\Delta_{fast} \sim \Delta_{fluctuation}$
Equilibration (quick iteration)	$\Delta t_{medium} \sim \Delta_{trade\ relationship}$
Trends (slow quasi-equilibrium variation)	$\Delta_{slow} \sim \Delta_{trend}$

Relative to these timescales, we might ask: what is the time needed to obtain a loan (for new money to be created)? How long does the loan last? What is the timescale of payback? So what is the supply of money available to flow from agent to agent and communicate transactions?

Several results from graph theory can help to shed light on the possibility for defining dynamical monetary equilibrium in such a network. One such result is the well-known Perron-Frobenius theorem, and its extension to directed graphs (with sources and sinks) in [BBCEM10]. These results prove that, for a graph with positive weights, there exists a principal eigenvector of the graph adjacency matrix, such that the fair weighted distribution of relative exchange values between the agents is represented in the normalized values of the eigenvector. An eigenvector equation assumes a linearity of a static adjacency matrix, however. In other words, it assumes that the money supply and the interactions (prices and trades) are basically static over the timescale the distribution needs to settle down and stabilize. In addition, one must allow for the possibility that there is no such simple separation of timescales. The potential even for static instabilities in such a network were shown in [BBCEM10]; for example, the presence of a source or a sink (a net importer or exporter with all of its neighbours, would result in the eventual draining of all money from the network, in an unsustainable manner.

The fact that the network is a directed graph, not an balanced equilibrium, implies that money flow is intrinsically unstable. The creation of money though debt is equally unstable to loops that act as positive feedback, e.g. a bank lends money in return for collateral financed by the bank itself along a different route.

Implicit in this result is a number of timescales that can lead to pitfalls in the naive application of the results. For instance, even if we believe that a simple infinitesimal linearization of the economy is a reasonable approach to approximation, the number of interactions that are needed to equilibrate a stable eigenvalue distribution $\Delta_{\text{trade relationship}}$ (analogous to Axelrod's game tournaments[Axe97, Axe84, Bur13b]). Applying this theorem conceals some details about dynamics that are not relevant to pure graph theory, and thus are not discussed there, but which are highly relevant to the dynamics of a real graph as a kind of cellular automation (see discussion and references in [Bur16b]).

We could easily use this to sum the balance of payments in a network of non-overlapping trading entities. As long as we can separate and trace the non-overlapping semantics of each entity it should not matter if agents a formally parts of other agents. This would only be a snapshot: timescales are implicit here as with any equilibrium

We shall not dwell further on this point here, except to mention that the same construction was used to define social trust and reputation in [BB06b].

9.4   SAVING BUFFERS OF MONEY THROUGHOUT A NETWORK

Saving for a rainy day sounds like common sense from the perspective of an agent expecting to be self-sufficient in a changing world. Only in a steady state, time independent, world would agents be immune to change. Having reserves of a dynamical resource allows one to weather storms and ride out hard times, i.e. to overcome both expected and unexpected obstacles. Buffers (redundancy) are a prerequisite for resilience to internal dynamics and external perturbations. Agents can build up buffers of money by absorbing, i.e. by releasing less than they absorb.

We cannot understand a money network without also understanding the network of things, goods, and services that attract money. Their appearance is not conserved. Taking money out of circulation might mean precipitating a situation in which there was insufficient money to buy new things, as they are manufactured, at the time of need. Markets could quickly become disabled if their natural network carrier (money) were shut down, or began to place quotas on data usage (how can you solve your problems, if you can't even talk to others?). This latter scenario is what happens during a depression, and during austerity. Agents become unable to buy their way out of economic obstacles, and the knock-on effects to flow can affect any connected parties in the network.

If monetary transactions were a Markov process, with no memory, or ability to accumulate savings, the economy would be like a perpetual game of billiards or pinball. Money payments would immediately 'bounce off' the receivers. When one agent moved its money, every other agent imposed upon would have to immediate pass the money on to another agent, or drop it forever. Since we have assumed that money is not dropped (officially or intentionally), that means money can never stop moving. But this would be a ridiculous scenario indeed: buying and selling would no longer be voluntary, they would be deterministically driven by the initial payment. The economy would ring like an echo chamber gone mad, and the agents within it would have no choice. This clearly defeats the purpose of money, which is to absorb transactions and allow local reservoirs of funds to accumulate and stabilize the money flows, which now fluctuate like the weather. Should a large payment become necessary (because a large bill arrives), the aggregation of savings could allow the agent to keep its promises, and continue to function.

The ability to loan money from a bank or other agent might also be a way to continue, assuming the agent has access. As we know, this access is not deterministically given in the real world. As agents save, the 'kinetic' money in circulation is not conserved on its own. Some of it disappears into pools and buffers inside agents, where it is turned into 'potential money', i.e. savings. If money can be exchanged between savings and flow, it seems clear that one dynamical possibility is that decisions by autonomous agents, locally, make this network unstable in one of two ways:

- All the money disappears into a few (rich) agents, so that none is left to exchange between others in the network.

- All the money floods out of the agents and there is too much for the network to carry, so some of it gets lost or the network is overloaded.

Intuitively, one expects that if any agent were allowed to accumulate too much money, this could be damaging to the network's stability. This is a scenario we can model with networks, using an eigenvalue problem.

9.5   FROM BALANCE OF PAYMENTS TO COLLAPSED EIGENSTATES

Money allows us to make the balance of payments distributed. We know the conditions for a sustainable network economy from graph theory[BBCEM10] (see figure 6.1). There are two approximate kinds of purchases:

1. Regular purchases of consumables: food, energy, raw materials, and other survival prerequisites.

2. Irregular opportunistic acquisitions: tools, televisions, holidays, armour plating, etc.

For regular purchases, acquired cooperatively from the network (i.e. rather than autonomous agent growing their own vegetables internally) a regular supply of money has to be available. This is like the pumping of blood around an organism. For any larger acquisitions, money has to be accumulated ready for the unexpected opportunities. Money savings enable this.

Any difference in the relative strengths of agents to absorb or emit money would lead to a network of queues, with backlogs to be processed at weaker locations. Even if we assume that agents are all capable of coping straightaway, money flows probabilistically in a network filled with loops and branching dependencies (see figure 6.1).

Under smooth predictable conditions, and within a single closed currency region, a probable equilibrium distribution of money would be determined by a simple eigenvalue problem, making use of the a knowledge of directed graphs. In a global multiple currency world, this is a simplistic picture, but not totally without merit in terms of understanding the stability of money, so it is worth a short discussion. For short enough periods of time, with conditions of sufficient stability (weak coupling) in an approximately closed system, one could imagine a collapse, or projection of states, onto a set of eigenstates that characterize the momentary distribution of money. Such an approach might be considered as part of payment clearing.

The Perron-Frobenius theorem, in graph theory, states that any strictly positive adjacency matrix representing a directed graph has a non-negative principal eigenvector whose distribution of components represents a solution to the weighted distribution of conserved flow in the graph[BBCEM10]. If we use a directed graph to represent the balance of payments between nearest neighbours in an economic network, then we can always make such a matrix positive by defining aggregating a matrix of promised transfers, whose direction points in the direction of positive transfer balance:

$$T_{ij} = \begin{cases} \alpha_O \left(\Pi_{ij}^{(+)} - \Pi_{ij}^{(-)} \right) & > 0 \\ \alpha_O \left(\Pi_{ji}^{(+)} - \Pi_{ji}^{(-)} \right) & < 0 \end{cases} \tag{9.1}$$

where i, j run over all economic agents, and $\alpha_O(\cdot)$ represents the assessment of a single possibly imaginary observer. We assume that the total number of agents is a constant population. These are obvious simplifications that can be addressed later. To make this consistent we have to also observe that payments can only be a fraction of existing savings (we ignore the possibility of borrowing for now).

$$T_{ij} \quad \propto \quad B_j \tag{9.2}$$

$$= \quad \Delta T_{ij} B_j \tag{9.3}$$

The new money received from incoming payments across the network is a sum over probabilistic promisers j and all product types a:

$$\Delta B_i = \sum_a \sum_j \Delta T_{ij}^{(a)}. \tag{9.4}$$

The form of this begins to take on the shape of well known stochastic problems in physics, for which there are many ingenious solution methods. So the total balance at each agent must be the previous balance plus the change:

$$B_i = (I + \Delta B_i) \quad = \quad \sum_a \sum_j (I_{ij} + \Delta T_{ij}) B_j \tag{9.5}$$

$$\equiv \quad M_{ij} B_j. \tag{9.6}$$

where I is the identity, or Kronecker delta, which leads to a self-consistent eigenvalue problem for the distribution of wealth (balance) relative to the promises.

$$\sum_j M_{ij} B_j = \lambda B_i. \tag{9.7}$$

Clearly, we can identify savings with the buffer of money accumulated at each agent:

$$\text{Savings of agent } A_i \propto B_i. \tag{9.8}$$

Moreover, although we have assumed only monetary promises in this example, by playing around with the matrix definitions, we can add in goods, services, and other kinds of promises that make society function, to discuss the limits on 'fair' or expeditious distribution.

The advantage of such a formulation (in spite of its representing a snapshot of stable population and money supply), is that it summarizes network effects, in a simple way, that are otherwise hard to visualize. The Perron-Frobenius theorem and its extensions tell us several things about the ideal case [BBCEM10]. There is a principal eigenvector B_i with non-negative values, that represent the level of savings available to agents that are in equilibrium with the states promises.

- A stable semantic trade and payment network can lead to a stable distribution of savings, over this short term epoch. However, in practice, this also assumes quite sober probability distributions for the stochastic behaviours. Long tailed, or so-called 'Black Swan' events will distort the distribution temporarily and take longer to converge.

- The economy can partition into a regions if and only if there is an imbalance of monetary flow in a single direction, or no flow at all.

- We do not need to know the precise promises or levels of trade to see that a region could stabilize, under constant conditions.

- If we only know a partial region of exchanges, boundary conditions can simulate the existence of exterior effects, such as foreign currencies, creation of new things, or destruction of old things[68].

- The role of banks as network hubs (see figure 6.1) makes them critical points for the redistribution of money in a network. If all payments go through banks, they can either act as fair calibrators or problematic bottlenecks, helping or hindering the sustainability of the network.

- If the savings can grow without limit, then an agent becomes a *money sink*, and can absorb the entire supply of money, leaving nothing for the other agents to exchange. The network is unstable to agents that are able to absorb too much. In practice, wealthy agents will grow at the expense of smaller agents: the larger they grow, the more they attract business, and the less impact payments have on their balance sheets. So, unless the promise landscape changes, the outcome is unstable to large agents grabbing all the money. This is dead-end for the economy, since other agents will not longer be able to acquire anything to make new consumables, and all the agents must perish!

What would society do if some agents took away all the money from the others? There seem to be three possibilities:

- More money can be distributed to those agents who have none.

- A new an separate economy could be built on a new currency for those who end up with nothing.

- Everyone returns to barter (essentially a collapse of society[Tai88]).

Of course, we need to be careful with eigenstate models. The self-consistent values assume a potentially infinite number of interactions to reach equilibrium. In practice, experience shows that only a few interactions are needed, and the distributions converge quite quickly. Nevertheless, a finite timescale for the equilibration is involved. There is no 'instantaneous' communication of influence, as proposed in the Efficient Market Hypothesis.

9.6 CLASSES OF MONEY

Money is sometimes classified in order of liquidity, from the most concrete tangible representations to complicated derived forms of value, using the "M" system (see figure 9.2)[MRT14].

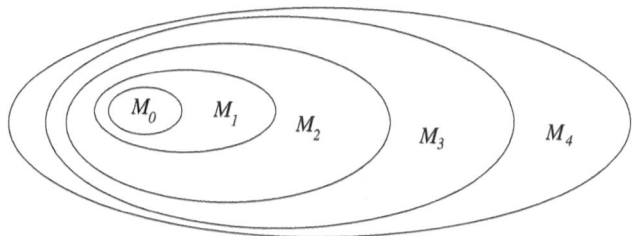

Figure 9.2: The money accessibility categories[Inv17b, Key30]. M_0 is cash in circulation (the most liquid form of money). M_1 is a superset of M_0, which adds public deposits at the bank. Then, the definitions become quite unclear. M_2 includes the aforementioned, and adds 'short term deposits at savings banks'. M_3 adds longs term savings and fund investments, etc. These categories are only schematic, and are used differently in different regions, e.g. for the Bank of England[MRT14],

- Notes and coins.

- M_0, notes coins and central bank reserves.

- M_1, includes M_0 and adds non-time (sight) deposits held by the non-bank private sector.

- M_2, M_1 adding retail time-deposits.

- Broad money includes all kinds.

A weighted summary of measures is published by the central Bank of England. These measures are, in principle, countable, subject to errors and unaccountable losses.

9.7 CAN ONE ASSESS THE TOTAL AMOUNT OF MONEY WITHIN AN ECONOMIC AREA?

Central banks purport to be able to keep track of the money they issue and create. Through regulatory insight, they can also receive reports from private banks on holdings and loans. In this sense, it is possible to count the promises made by regulated financial stores of money. By definition, banks cannot define the amount of fraudulent money proxies in circulation, whether counterfeit notes or incorrectly accounted payments on ledgers.

If we were to reset the financial system, it would be plausible to track modern money quite accurately. However, since we did not start this until there was already money in circulation, there must remain doubt about the legitimacy of promises made by different agents.

9.8 INTERIOR AND EXTERIOR USES OF MONEY

Money is created by banks who locally centralize their ledgers for consistency, through clearing. When money is transferred to another bank, the consistency would be lost without banks keeping regulatory promises or standards of behaviour. The same is true of monetary movements inside versus outside companies and other organizations. In general, the trusted status of money between agents is different from the trust within an agent.

Assumption 14 (Interior and exterior trust). *The trust relationship inside and outside the boundary of a (super)agent is different. Outside every calibration is made peer to peer, longitudinally, or as a cognitive learning relationship. Inside, calibrations are more likely to be made on a common basis by default trust, using ensemble learning.*

The agent boundary is clearly important, because it defines a region presumed to consist of *similar promises*. What happens when we cross over into a different system of promises, even simply a different set of units? Money has to be converted by buying/selling because the relative conversion rates are not constant, and the semantics of the monies might differ too. There may be spacetime-dependent relativity between currencies.

Example 62. *Consider some humorous examples, based on Einsteinian relativity. Imagine paying for satellite communication time in a special satellite currency. Satellites in high orbit experience a different gravitational field, and atomic clocks run at a different rate there. Whose clock do we use to measure the time paid for, if exchanging satellite money for Earth money? Or imagine interplanetary traders approaching each other at close to the speed of light, wanting to buy a yard of ale or cloth. The buyer and seller measure the yard quite differently. Whose assessment of the length of ale or cloth is assessed as correct? in IT systems, at high speeds, and micro-payments, these details are not completely irrelevant.*

If ledger money leaves a single ledger, its integrity cannot automatically be trusted. There is nothing to stop a non-regulated or non-trustworthy agent from inventing more money than it should, or finding ways to add or subtract from transactions. Could or should an American bank create Euros at its branch in the Cayman Islands? The common assumption is that kind of behaviour this is regulated i) by 'the market' (this would have to be on a timescale much larger than that of a suspicious transaction), and ii) by impartial agencies (this can be done on a regular timescale, such as each day)[69].

As we have noted, in section 6.11, semantic constraints on money proxies could prevent such fraud in future, but could also lead to discriminatory acceptance of different forms of money (new kinds of fraud). An alternative to this, worthy of future investigation, might be to separate money by scale, so that formally different currencies are used for the transfer of money between institutions at any given scale. This is something analogous to an Internet firewall model.

Definition 102 (Interior money (endogenous)). *Money that is trusted within the bounds of a single agent, for interior use. Note that what is interior and exterior is scale and agent dependent.*

Definition 103 (Exterior money (exogenous)). *Money that is trusted between agents, for exterior exchange. Note that what is interior and exterior is scale and agent dependent.*

Once money leaves the superagent boundary, it is transformed into a new currency, with an impartial regulator, so that money for buying from other agents is different from money for buying inside an agent. International trade could have a single common currency, different from any country's local currency.

In practice, exterior money already exists, by any other name, at the scale of nation states is often tied to bonds, deeds, and other financial assets, but fluctuating currency rates leads to complexities that could be simplified by a scaled approach the monetary promises. Even banks use bonds, treasury bills, and other securities to clear temporary transactions. The problem is that the selection of a mechanism for monetary transfer is not handled by concerns about information integrity, but rather by commercial profiteering and the timescale over which money is required to be transferred to clear a transaction. Time, again, is at the heart of the definition of semantic validity and dynamical stability.

9.9   MICROCURRENCIES

The separation of concerns is practically a hallowed principle in information science. In computer science, for instance, this is often expressed by preferring local variables over global variables. It is a principle that avoids dynamical and semantic interference of information. If one applies the idea of separation to money, it suggests that 'local' money would be preferable to 'global' money, where local and global refer to the scope of its applicability. In the past, scope has been based on geography, but it could be based on any semantic distinction. Taking this to an extreme, why have one global currency for all kinds of things? What if every economic relationship had its own kind of money? What would be the pros and cons? We have shown that there is conflict of interest in giving money greater semantics.

The separation of human resources, effected by the labelling of things by an owner, has wrought havoc with fair distribution throughout history[Har11, Suz17]. Ownership partitions things and money partially reconnects them, but it does so preferentially to the agents who happen to have it agents. Money's weak semantics have not been able to compensate for the effects of social partitions effectively. So is it really a good idea to create even more walled gardens, with even more kinds of wall, in a network economy? The lessons of free markets versus protectionism no doubt play into the pool of evidence here. We shall not attempt to offer an answer there. However, it is worth discussing whether it matters *how* we draw the boundaries around special regions (i.e. by what criteria). The goals of society were totally changed by the concept of private ownership, and materialism, when money was freed from bonds of the gold standard, for instance. We should not discount the possibility that they could be redrawn once again, as the technologies and costs of distribution change in the future. For example, the idea of globalization is meaningless if all countries have free access to the same goods and services through their home replicator.

Part of money's appeal is its equalizing character. However, money is no more of an equalizer of opportunity than is language: it is how it is used in the hands of agents able to facilitate change that matters to society. If you don't speak the language, having a large bag of words doesn't help you. Cutting down noise is just as important in the fidelity of communication as is sufficient bandwidth. So separating off truly independent concerns may actually improve access to money.

The only label traditional money carries at present is an amount. Thus traditional money can discriminate only on amounts (see section 7.16). Could this be the reason for a rich-poor divide? If we introduced different currencies for each agency and scale, would there be other divides (other than the obvious ones within sovereign regions)? A separation of trust relationships could be a way to restore the bond between money and society. We can already see the rise interior monies in businesses, including loyalty point systems. Recall:

- Airline miles—moneylike tokens.

- Coffee cards—tokens, say 1/10th of a cup of any coffee.

- Buy one get one free.

- Discount on your next purchase of 'commodity'..

- Petrol stamps (Green shield, Coop)

- WeChat.

- Paypal.

- BitCoin.

- World of Warcraft money.

Loyalty is closely related to trust; building currencies based on loyalty reintroduces the idea of familial and tribal bonds, preferences, and potentially warlike conflicts between groups that the introduction of trusted institutions helped to bring minimize in civil society. These private currencies can bypass taxation, currently, which could become increasingly unsettling for governments as the size and scope of the currencies grows. Many of the companies (e.g. airlines) have sufficient size and financial stability, across multiple commodities and resources to act as a major "state" powers. It is sometimes said that corporations are the new nation states, and these monetary forms suggest that the future of sovereignty will become virtualized with respect the its traditional geographic borders.

A possible compromise for monetary semantics would be to insist that all forms of currency were regulable according to the central rules of that currency. These would at least be impartial semantics, advertised in the common interest, rather than secretly discriminatory. Discriminations of currency exchange would then be directly aligned with the politics of each region, and users could make up their own minds about whether to use them or not. We believe it is an empirical fact that a single set of rules and policies, for all cases, never leads to a working system that is considered fair or able to adapt to dynamical context.

One possible observation could be that the purpose of a microcurrency is not to tie a purchase to a specific good, but to a specific community. What happens within a community stays in the community. This aligns, at least, with trust relationships, in the sense of Dunbar and Axelrod[Dun96, ZSHD04, Axe97]. The key problem to be solved by society and its conventions is how we exchange these currencies. Exchange itself can be automated and impartial, and thus a common exchange currency can be strongly regulated without needing to offend human eyes. If the software to manage it could also be authorized and monitored by all parties, fairness could be policed in an open forum, reducing the risk of fraud.

Example 63. *China's third party online payment systems had total mobile payment transactions worth 820 billion USD in the first quarter of 2017, led by Alipay (34.7%), then UnionPay and TenCent Finance (WeChat) [CIW17]. The rise of these cashless payments in China has made Chinese cities the 'smartest' of the smart cities already, opening platform possibilities for sophisticated online information semantics.*

9.10 NETWORK COVERAGE OF MONEY

Just as the network coverage of telephone and Internet networks is patchy and uneven, so they coverage of money networks, issued by different providers, is patchy and uneven. Certain kinds of semantically neutral money, belonging to a single currency, could be simplified by having only a single ledger and loan mechanism rather than the multiple ledgers of multiple banks in the current system. Indeed, this is already the case for private microcurrencies, such as air miles. However, this approach limits the semantics of a monetary network, so it cannot be a solution for 'smart' features. The tagging or earmarking of money is likely to be   an important function that prevents misuse and misallocation of funds. Tagged money could have rate (velocity) controls built into it, preventing bubbles by oversupply or overbidding. With stronger semantics, policies for allocation could replace mere auction bidding, e.g. First Come First Served, and quotas on buying. Such semantics are common in other network information systems, but are conspicuously absent in the financial system.

Context is also important in accessing money. Access to money is highly non-linear today. The availability and semantics of money, relative to its users, may be very different depending on whether agents are rich (luxuriating) or poor (subsisting). As we show in section 9.5, the network is biased in terms of pre-existing wealth. The ability to use money to pay for transactions is also restricted—the network is not free or unbiased.

Promise Theory underlines this point with its explicit focus on (+) and (-) promises: it is not enough to be able to promise money

Example 64 (Do you take credit cards?). *Many credit card companies famously boasted about the acceptance of their cards over the years, and some countries are almost entirely cashless societies—yet strange anomalies persist. Taxi drivers in many of the most advanced countries refuse everything but cash[70]. Conversely, many cities with payment systems for subway, metro, bus, public transportation (e.g. Oyster, Octopus, and YoYo cards) may accept payments using those 'private' currencies at kiosks, cinemas, and other vendors. This is highly convenient for users, but adds a burden of cooperative infrastructure to the providers. Credit cards often pass on fees to the vendors to conceal them from end-users, which makes them less popular. The card brands (VISA, Mastercard, etc) also issue both credit and debit cards—and some vendors will take one or the other, but not both! We end up with the peculiar patchwork of agents who will and will not accept different kinds of money—creating uncertainty and trust issues for users.*

Example 65 (Airline miles). *The major airline groups share loyalty card point systems, allowing users to accumulate a microcurrency for 'payments in kind'. The points can be used to purchase goods within the network of partners, including food, hotel nights, car rentals, and more. The closed networks are semantically restricted, but share common semantics of travel-related matters. So, although the money is less fungible than general purpose money, it has a context-specific fungibility nonetheless.*

Promise Theory's law of intermediate agents tells us that promises are not transitive and do not automatically extend through middle-men or third parties. This is an important issue when agents act as transducers between different parties in a transaction. Third parties break transactional semantics.

Example 66 (Middle-men and escrow). *A customer orders an item on a foreign website, using a payment agent who will perform a currency conversion. After a short processing delay, the foreign vendor replies that they cannot promise to deliver the goods after all, and the transaction fails to commit. The money is refunded, but the payment agent returns less money than the customer committed because the currency conversion rate has changed. Is this fair? In this case, the payment agent is charging the customer before goods are received and passing the amount to the foreign vendor in their currency. The customer ends up paying for the inconvenience. A natural solution is to settle by escrow: i.e. the payment agent holds a fixed amount of money, which is only released once the goods are delivered—failure to do so releases that fixed amount of money. This has the proper transactional semantics expected for money payments.*

As money separates into distinct types, which cannot be interconverted freely, new markets have to come about, and system integrators will offer unification schemes—just as content providers integrate semantically distinct (branded) services of different content creators.

Example 67 (Content and syndication). *Television and movie producers limit the viewing of their content to specific private channels and users may have to subscribe to multiple redundant services in order to view their content. This is inconvenient for uses, so there is an opening for system integrators to collect all channels into a single offering.*

This phenomenon also happens for microcurrencies of all kinds. The airline miles partnerships are an example of this for private loyalty points. The Euro is an example of such a system integration for the European countries.

Integrators will also seek to distinguish themselves until they yield to the forces of commoditization. In all semantic systems, this leads to a growing pyramidal hierarchy of integrations and diversification, which helps to sustain the growth model of the larger economic network. Olsen points out in his classic book *The Logic of Collective Action* that the traditional idea that large and small groups are equally effective in maintaining influence and coherence is wrong[Ols71]. Promise Theory predicts that dynamical (quantitative) discriminators will always be the arbiter of what is viable at any scale; however, given that basic viability—for smaller systems—the voluntary association of agents in a collective depends on the semantics of their common purpose up to a certain size, beyond which the cost of distinguishability becomes too expensive.

Introducing a private currency channel does not fully isolate a collective from the influences of external agents, but it adds a level of semantic 'firewalling', which allows the currency to discriminate between its users, by adding a semantic boundary. This is a simple form of control which can be exerted by any network provider, on the basis of distinguishability. All collectives are driven by common interests on some level. If those interests are cheap enough, they can be provided to everyone equally (as a commodity). Scale eliminates semantic discriminators, as one expects of statistical systems.

The role of integrators exposes the imposed inconvenience of such control for users, who may be willing to pay to dissolve it; meanwhile, the firewalling of context-specific currencies exposes the convenience and value of information to maintain distinctions in any system with 'smart semantics'. The scaling argument suggests that there will always be a minimum and maximum size for any agent collective, beyond which certain promises are not assessed as viable. The emergence of universal scaling occurs when the alternative roles are more or less equally represented up to a scale factor at each scale of integration.

9.11 MONEY AS A LEVER FOR CONTROL

Management of money—in terms of its supply, access to lending, and the right to create or destroy it—was exposed as a key mechanism for economic policy in the twentieth century, championed in particular by the work of Keynes[Key30, Key64, Til07]. Though it remained ignored for decades, the network role of money is now becoming an increasingly important picture for an information age.

Classical economics is built on many simple quasi-deterministic (time-independent) scaling relationships, which derive from an era in which economists looked to classical physics (in particular to thermodynamics) to model their ideas[71]. But the twentieth century was also the one in which such determinism was completely undermined as the foundation of behaviour in complex systems. This revolution is only beginning to impact on economic theory, as both the semantics and dynamics of money, in all its forms, are finally being recognized across the network of multiscale systems.

One such argument concerns the relationship between the money supply, the supply of goods and price inflation. Suppose there is a fixed amount of money in the money supply μ, and there is a fixed amount of stuff to buy $|T|$. Then, a simple interpretation of price P is to assume:

$$|T| \times P = \mu, \tag{9.9}$$

i.e. the supply of goods multiplied by the price is shared between all the money tokens. This predicts that price scales roughly as

$$\text{Price} \sim \frac{\text{Supply of money}}{\text{Supply of things}}. \tag{9.10}$$

So, naively, if we increase the amount of money in the economy, the prices will increase. Conversely, if we flood the market with goods, their price will fall. The equation has a comforting determinism about it, and it provides a simple lever for managing inflation by managing the money supply—but the reasoning and the conclusions are wrong on a number of levels.

- The picture assumes a basically static economy, and any changes are transmitted instantaneously and rigidly to all agents.

- Not all money is used to buy things. Money can also be saved, even hoarded, and later dumped into circulation to manipulate prices by any agent.

- The proportionality confuses how buyers determine the amount they are willing to pay ('the market value') with the actual price of a thing. Things are not instantaneously valued and priced by consumers according to economic arguments about goods supply and the availability of money. There is a finite time-lag, or hysteresis, in all processes is the key omission in this thinking. The economy is a dynamical system—the local money supply is a dynamical system[72].

Demand and supply do not move in lockstep—there is hysteresis and selectivity based on semantics. A key mistake in economics is to treat all things as commodities, forced into marginal prices by competition. Agents in an economic network don't live in an instantaneous present–they are constantly trying to guess the future, and fluctuations in buying power depend on guesses about inflation, as there is always a time-lag between a loan and repayment, or goods pricing and payment date. Small or large deviations from predicted expectation get amplified over time and may not even out as the assumption of a static economy would have it. Before we know it, agents have uncontrolled debt, or price bubbles emerge from speculation. So the simple guiding relationships can only ever be true as the most indiscriminate average. There can still be bubbles and crashes in individual markets whose average effects cancel in econometric terms.

The other essential gradients, which relate to the rates of change over time, are the interest rates. These both guide policy and are the single lever for trying to adjust behaviour. The relationship between prices, average inflation, money, and interest rates has been an interesting lesson since the 2008 financial crisis. Economists led us to believe for many years that the deterministic consequence of too much money floating around was simply inflation, but that turned out to be wrong: in some areas, it also lead to negative interest rates, or a combination of both. The latter attempts to use the leverage that interest has to guide behaviour, assuming agents to be rational—though it remains to be seen if it can curb the worst excesses of speculation. Keynes was keen to emphasize an economy could only be stimulated in a healthy way by investment, not by simple spending or lending[Til07]. The rise of financial speculation and increasingly devious networks of debt refinancing has contributed to amplify instabilities.

Returning to the issue of ownership of money, we can now draw attention to the role of the *money provider* (usually a central bank) in exerting influence over its users. Promise Theory exposes the similarity between a monetary network and any other kind of membership organization, whether it be a sports team, a fan-club, freemasonry, or national citizenship. The providers of the network have the ultimate power to manipulate and game the rules. We have seen this in other key network hegemonies such as the telecom companies and the energy companies to steer markets. Ultimately all such networks have been subject to scrutiny and regulation, as the 'invisible hand' was only helping itself.

Today, the central banks, card companies, and electronic payment companies use (for the most part) a standard set of sovereign currencies, belonging to the nations who subscribe to them, and promise to exchange these for foreign payments, by some usually opaque process. In effect, by adding fees to these distinguishable currencies, such integrators can exert influence and even rule the effective usage of the currencies as if they were an independent currency. This indicates the way in which an information technology can bias an economic network in a powerful way.

There are clearly limits to the extent to which a monetary integrator can manipulate its members in the use of currencies. Agents remain autonomous, even when they may be induced to cooperate by external pressures. By making conditional promises to its members, an integrator can impose a 'closed shop', like a workers union, in order to align with certain political or economic goals. If the value of the integration is sufficiently high, and unique, the integrator can thus impose conditions on its members, blurring the lines between what is voluntary and obligatory[73].

The value-added service an integrator provides includes the ability to shield its members from the specific semantics of different currencies. However, as long as money markets set the rates for exchange between currencies, the extent to which an integrator can mimic a completely independent currency may depend on its ability to sacrifice buffer reserves. An integrator could, nonetheless, employ a long-term strategy of extracting payment from its members in order to build the reserves for later strategic depletion in order to impose its own private agenda—just as credit card companies do.

Each integrator acquires some power to exert control over its users, because it controls access to its valued service. Where there are common interests, there is often a systematic exploitation of individuals by the group, as long as the cost of maintaining the coherence of the group is low enough to be viable. Conversely, small agents may be able to play the role of free-riders[Ols71].

Example 68 (Semantic money). *The distinction between smart services and the money they use is being eroded in the smart device era. Today, Chinese users can dial a taxi through a Chinese electronic currency platform and get a Chinese-speaking taxi driver. Context tailored services, based on microcurrency promises drive not only loyalty but value added service.*

The latter example is particularly interesting as it offers a way in which the capitalist economy could be used to undermine national borders and attempts to control the economy. Governments have only trust in compliance with law as a way to tax services as money relationships become increasingly opaque.

Example 69 (Control and scale dependence). *In a purely quantitative (scale-free) world, one expects a system to be characterized by its dimensionless ratios, e.g. $8/4 = 4/2 = 2/1$. But, in a discrete system of agents with semantics, this need not be true. The total size of a superagent collective may select a unique set of capabilities.*

Example 70 (Monetary warfare). *The network provider of a particular currency, as a Trusted Third Party, has considerable power to control the agents who subscribe to its services. One can imagine a scenario in which an alternative currency network is introduced alongside a country's main central bank currency, in direct competition with it. Over time, this currency could completely supplant the original currency. If the currency belongs to a foreign political power, this has the semantics of a political invasion, and the foreign power could then hold the users to ransom. Of course, this doesn't only apply to countries—any large actor could use this method of subvert the normal political system of a region.*

This example has been one argument for the 'democratization' (sic) of money using cryptocurrencies (see chapter 11) to diffuse the strict centralization of control.

CHAPTER 10

MACROECONOMIC MODELLING

'In the twentieth century, a totally different view of public debt emerged, based on
the conviction that debt could serve as an instrument of policy, aimed at raising
public spending and redistributing wealth for the benefit of the least well-off
members of society. The difference between these two views is fairly simple: in
the nineteenth century, lenders were handsomely reimbursed, thereby increasing
private wealth; in the twentieth century, debt was drowned by inflation and repaid
with money of decreasing value. In practice, this allowed deficits to be financed
by those who had lent money to the state, and taxes did not have to be raised by
an equivalent amount. This "progressive" view of public debt retains its hold on
many minds today, even though inflation has long since declined to a rate nor
much above the nineteenth century's, and the distributional effects are relatively
obscure.' [Pik14]
–T. Piketty

10.1 SCALES AND ECONOMICS

This is not a book about modelling the economy, yet we hear far more about 'the
economy', as an entity, than we hear about money. Is there a relationship between the
two? We expect *scale* to play a role. *Macroeconomics* is the story we hear on the business
news; its central concepts (GDP, employment, inflation, interest rates, stock markets, etc)
ring like the familiar slogans of a branding campaign, but remain mysterious to many.
Microeconomics is our day to day experience of paying for things with money, and the
moral compasses of wages, purchases, debts, and consumerism to guide us. The link
between the two, however, is mysterious, highly politicized, and filled with riddles.

When there is unemployment, scarcity, or prices rise, we blame 'The Economy',

195

but what is this capitalized economy, pun or no pun? Is it the macro-economy, or is our personal micro-economy to blame? What do broad sociological changes like employment have to do with how much interest we should have to pay on a mortgage? Our personal experience of economic trouble might be that of not having enough money to pay bills, but does that view scale to the level of a nation state, or the globe, or could politicians simply print more money?

Many of the narratives we have about economics come from experiences that are not directly relevant to the matters where we hope to apply them. Economists too seem to suffer from an inability to separate the mechanics of the economy from its intended purpose. That purpose is deeply political, so we could never fully separate the machinations of money from policy. Nevertheless, these notes are an attempt to understand the independent influences and where they can and cannot be separated—to underline some of the implications of muddling concepts, and see where the current narratives are flawed.

10.1.1   CHARACTERISTICS OF MICRO- AND MACROECONOMICS

Microeconomics is a kinetic theory of the economy, ballistic and transactional, with payments for goods and services firing back and forth like billiards or particle physics. What about the macro-economy? As economic studies developed, during the industrial revolution, its philosophers went to some lengths to model the capitalist macro-economy on the successes of the macroscopic physics of the day, namely the equilibrium thermo-dynamics of heat engines. This was an intentional design, based on 'physics envy', and it persists today[74] .

Mismatches of scale are a constant dilemma for the description and governance of systems. We witness how characterizations, laws, regulations, and policy set for the average case fail to address the needs of individual cases. When one treats all cases in aggregation, the result may apply to none, and average concerns risk riding roughshod over individual concerns.

10.1.2   POLITICS AND ECONOMICS

We expect intent, politics, and moral justice play a large role in our ideas and perceptions about the economy. For some, economics should describe a system supposed to be for the benefit of society; for others, it has become a detached and impersonal game of accumulation and loss, to be mastered by 'winners', a game for social status[75]. Economists themselves have frequently sided with political positions, and shaped their conclusions with deliberate political slants, bending the theory to fit their opinions rather than facts[Fri02, Hay44], only later to be uncovered[Min82, Var15].

Societies in the developed world are very much designed for those with money: and those with plenty do not need to encumber themselves too much with matters of society and justice, because money insulates them from many encumbrances, and they can operate with a high degree of autonomy—decoupled from mundane issues. Social and legal protections are needed mainly by those who are not of independent means, who experience a less well-functioning social infrastructure, and who naturally search for answers in its founding principles. If we don't have money, then someone else must have it, which leads to questions of equality and justice. We are suspicious of a system that leads to too much inequality.

Piketty has argued that significant wealth cannot easily be earned as the product of individual work—rather it is accumulated as 'capital' over generations, often fortuitously, and passed on through family dynasties by inheritance. It's then amplified by rent collection on property[Pik14]. Thus the ability to grow one's wealth significantly depends on what 'rentable' assets one has. Such rentable assets are what we know as 'capital'.

Whatever 'The Economy' might be to any one of us, it impacts on the lives of the ordinary and the extraordinary alike. In practice, what we hear about—what governments estimate is 'national average output', aggregated for an entire country, one year at a time. This makes for interesting statistics on social progress, but it has little to do with any person's experiences. By contrast, governments and central banks often imply that their decisions and actions can 'fix' the economy, when it seems to go awry (as if by sheer political will). The problem with statistical models is that, if the macro-economy is about an average condition for everyone, then it actually refers to no one. The verdict is that such political interventions are often haphazard, and have often been misguided[Min, Kee11, Kru08, Var15].

10.2 MACRO-MODELS

Based on the Newtonian tradition, economists studying macroeconomic models have been drawn to smooth differentiable functions to model changes in aggregate variables, assuming the kind of stability that was the basis for the engines of the industrial revolution. As one might expect, this had some success in describing large scale aggregate behaviours on timescales at which changes to averages could be considered smooth (years and decades).

One macroeconomic phenomenon, which has been modelled in this way, is the cyclic pattern of 'boom and bust' reputed in capitalist economies. During a boom the economy grows, jobs are plentiful and the market brings profits to investors. In a bust the economy shrinks, people lose their jobs, and investors lose money. Economic models, such as Goodwin models, and their extensions as 'Minsky models' by Keen[Kee95] and followers[GH15, GH16], are differential formulations of economic cycles in a closed network. These cycles of employment, earnings, consumption, capital investment, and rent collection, compete with one another to bring about cycles of 'boom and bust', though the reasons for such cycles are not without some controversy[76].

The accounts of economics by Minsky are particularly useful as a basis for describing a multi-scaled view of economic activity[Min82]. His descriptions are both detailed in capturing real world details, and focus on the mechanics rather than the politics of phenomena. Using promise theory to analyze Minsky's decomposition of the macro-economy[Min] may help to reveal some of the structural dependencies, and potential failure modes, of our present day economic structures.

10.2.1 PROMISE-BASED MACRO-MODELLING?

The elimination of small scale semantics from present day models of the macro-economy, suggests that semantics are not important on a longer timescale. However, the choices, checks, and balances that are used to steer the economy are intentional and span years too. It seems plausible that we can identify a network of economic activity as a kind of semantic process[Bur14, Bur15, Bur16b], and even as a cognitive system (whose sensory apparatuses are yet to be defined).

Promise theory asks us to consider: who and what are the key agents in a network of interactions, and what promises do they make to one another?

- Agents at a microscopic level could be individual persons, households or firms, interacting amongst themselves.

- Agents at the macro level may include nation states, governments, and central banks. The promises macroscopic agents make at the macro level cannot directly or immediately influence microscopic scales (just as weather does not directly change people's behaviours to any large extent under normal conditions), except as an effective (non-linear) boundary condition; but, with a powerful information technology, multi-scale agents could self-govern by detailed balance, i.e. by brute force (like a Maxwell's daemon) countering every debt with a payment.

In aggregation, agents may accumulate into certain classes, playing particular roles, but we should be cautious about inferring smooth statistical properties to these bulk quantities, as remarked in section 10.3.

10.2.2 MACRO TERMINOLOGY

Let's briefly review some terminology used by macro-economists in terms of the language of this book[77]. Promise Theory considers agents, working together in clusters. Some agents can act on behalf of others in a systemic role. Big government can redistribute money, for example, to prevent it from falling out of circulation in wells that halt economic activity, but the political machinations can also focus and amplify intent, concentrating demand for something by biasing intent.

The semantics of any large coherent influence, by a large company or a government can bias a price profile. Timescales are similarly affected by the formation of 'superagent' institutions, like firms and government—these increase the lifetime of agents by spanning multiple human lifetimes, and taking on an independent purpose. Large institutions are needed to carry out large projects, as these are the only ones that can raise sufficient financing. Thus large and small agents play qualitatively different roles.

Money moves from account to account, through the monetary network, conditional on the account holder's (A_i) authorization:

$$\text{Account}_1 \xrightarrow{+\mu|A_i} \text{Account}_2 \tag{10.1}$$

Occasionally cash might be exchanged, in which case there are no account holders, and individual holders H_i are involved:

$$H_1 \xrightarrow{+\mu} \blacksquare H_2. \tag{10.2}$$

Figure 10.1 sketches how the approximate separation of concerns in economic semantics broadly follows implicit spacetime scales, where space refers to aggregation of instances, and time refers to accumulation of duration.

The scaling of payments from micro to macro levels involves a change of perspective:

$$\text{transactions} \xrightarrow{\text{aggregation}} \text{flows} \xrightarrow{\text{duration}} \text{investments} \tag{10.3}$$

$$\text{things} \xrightarrow{\text{duration of promise}} \text{capital} \tag{10.4}$$

Thus, economists will speak of spending and investment (which a physicist might call fast and slow spending, or short term and long term money interactions). Economists speak of commodity goods and capital, where a physicist might only note the timescale over which these goods are held and are used up[78] . Capital items (long term goods) tend to be more expensive purchases, and they tend to be usable for processes that can lead to (rent collection or manufacturing), whereas consumer goods tend to be eaten or used quickly (and although they fuel all our activity, we choose not to see them as wealth creating).

The binding of money in potential ('capital') rather than kinetic ('liquid') form is a key idea. The term *liquidity* is used to mean how easily a form of money can change hands, by analogy to flow. The terms free (kinetic) or bound (potential) are commonly used in natural sciences. Money might be viewed as being locked into a purchase, as if it

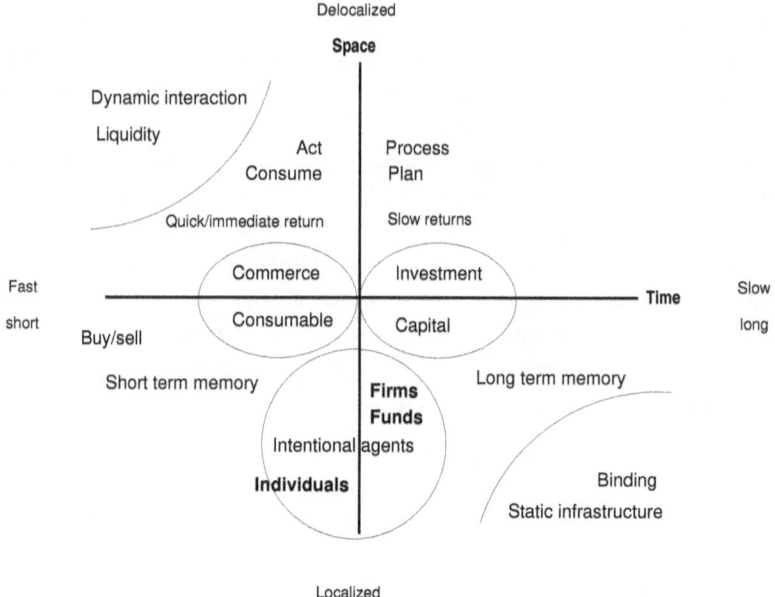

Figure 10.1: The interaction scales for economic processes can be organized by spacetime scales. The traditional separation into capital investment and commodity consumption is not significantly different in semantics, but different in timescale.

were a chemical compound waiting to be released by some suitable reaction (the analogy is not important, but the scaling is). Cash is free to be spent and is thus liquid. Indeed, the main factor distinguishing *capital* from *commodity goods* seems to be a timescale for the keeping of its promises. Capital is an illiquid asset, whose resale price decays slowly, perhaps after appreciating in a specialist market, while a commodity's life cycle is short and liquid, with therefore little opportunity for resale or change of price. Luxury goods lie somewhere in between these cases: if not immediately consumed, they might be resold (other than wholesale/retail distribution).

The term Return On Investment (ROI) is widely used in business to refer to an amount of rents or profits that can be extracted from a capital asset. This is different from what the immediate resale value might be. So what is the profit potential of a property would come from renting out the property, or from holding on to the property in a rising market.

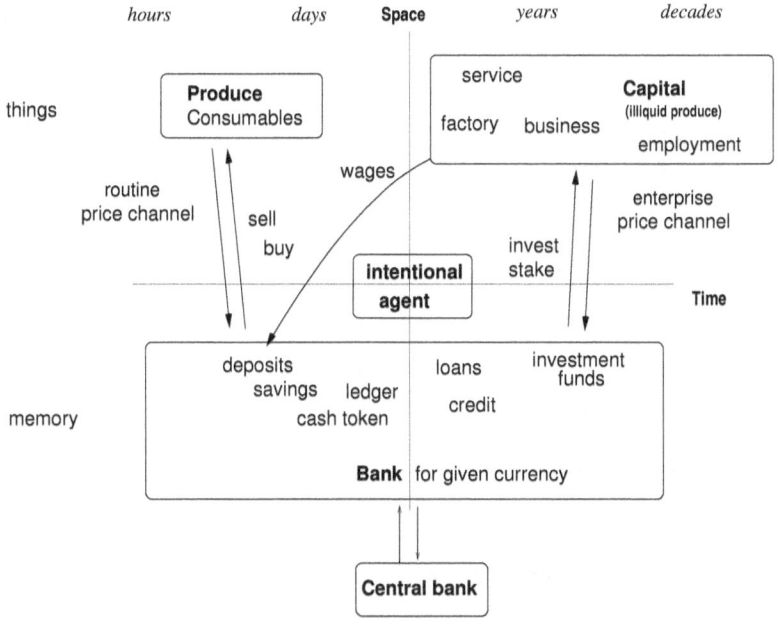

Figure 10.2: The flow of money in the economy does not pass through humans, except for cash transactions (which are increasingly uncommon for all but small daily exchanges). Humans are only conditional authorizers of transactions that happen between different ledgers. The connection between money and property is quite tenuous, and cannot be maintained naturally without the machinery of legal convention.

10.2.3 INFLATION

Inflation is a key parameter for developing a scaled view of money over time. It links the supply of money to the assessments agents make about its purchasing power (loosely termed 'value'). There are two kinds of inflation[Min82]: i) when wages rise at a slower rate than prices (on average), but they keep pace in a more or less linear fashion, and ii) when rising prices induce rising wages, in an unstable spiral of positive feedback[79]. Prices may increase if the supply of a certain thing (or its indistinguishable alternatives) is short[80].

Under 'normal' conditions, inflation is a second-order, relatively slow process, compared to buying and selling. This is not true under conditions of hyper-inflation, during which the economy is already on an unstable trajectory. *Investment*, or injecting money into a market may impart an inflationary push, as relative wage rates reflect market demands. Purchasing power may not change if one good rises and another falls in price. The net result could be the same, so there is an aggregation effect over many individuals that gets converted into entropy with scaling—which is an argument in favour of statistical modelling[81].

When new money is instantiated by banks, along with debt, and that money is repaid with (usually positive) interest, meaning that more is repaid than was created. So more money in circulation is needed in advance to enable repayment of positive interest. How does the money supply increase to accommodate the repayment of inflation and interest? Inflation is generally assumed to be the mechanism implied to cover that. Repayment of debts leads to a contraction of the money supply however, so it is not in anyone's 'interest' to repay too many debts in such a system. The creation of more and more assets, refinancing, and bond instruments essentially have to finance positive interest repayments. Another important aspect of this, which is barely modelled, is perishability of assets. What effect does depreciation and obsolescence have?

The echoes of debts are held in this memory of bonds and refinancing. Inflation of prices, and thence the money supply, and subsequent earning may gradually erode debt, so that no one actually repays the initial promised burden of borrowing. However, this is now a complex matter given the use of negative interest rates, especially in Europe. Having more money in circulation, when interest rates are positive, need not create inflation if the velocity of monetary circulation decreases; and negative interest rates need not lead to inflation if the contraction in the amount of circulating money is compensated for by the increase in the velocity of the money. It seems likely that both positive and negative interest rates may lead to inflation: in the former case by trying to offset the burden of debt, and in the latter case by being forced into spending and triggering a market expansion. Inflation can thus a destabilizing influence in all cases.

10.3　Goodwin Models of the Economy

Following the tradition imported from Newtonian physics, economic dynamics are usually modelled as smooth macroscopic drifts using differential rate equations. This could easily be criticized as choosing models to fit the tools, rather than vice versa. The scales over which smooth differentiable functions can be defined for economics suggest that such models might describe changes over the past several decades, rather than the coming weeks. So, while they have some interest for the purpose of seeing patterns and cycles in the long term economy, given certain assumptions of invariance, differential models have probably little role to play in predicting the daily concerns of ordinary economic participants.

Goodwin first tried to model the interaction between consumption and investment scenario (a so-called class struggle model) in a simple local 'mean field theory' at the macroeconomic level, in order to investigate the normal modes of oscillation between

GH	YBY	SK	Units	Meaning/Semantics
N	N_w	N	[1]	Population
ℓ	N_w	L	[1]	Workforce $\propto e^{\beta t}$
λ	λ_w	$\lambda = L/N$	[1]	employment level/rate
p	P	-	$[\mu]$	Price level $\propto e^{\gamma t}$
a	θ_w	a	[1]	Productivity per population $\propto e^{\alpha t}$
W	W	W	$[\mu/t]$	Wages
Y	Y	Y	$[\mu/t]$	Agent output
Π	R	Π	$[\mu/t]$	Return on Investment / Profit
C	C	-	$[\mu/t]$	Consumer consumption
I	I	$I?$	$[\mu/t]$	Capital Investment rate
K	-	K	$[\mu]$	Capital held
pI	B	-	$[\mu/t]$	Inter-agent trading/purchases
L	-	D	$[\mu]$	Debt level (loan) of agent
D	-	-	$[\mu]$	Bank deposits of agent
r	-	r	[1]	Interest rate
$w = W/\ell$	w	$w = W/L$	$[\mu]$	wage level per capita
$\Phi(\lambda) = \dot{w}/w$	-	$w(\lambda) = \dot{w}/w$	$[t^{-1}]$	wage rate

Table 10.1: Some literature conventional notations compared. In this paper, we follow the notations of GH[GH15, GH16].

investment in capital and consumption of consumables, and their stable equilibrium states of these modes. The result was a set of coupled differential equations, based on the assumption of exponential growth of population and output[BYLMKG17]. The model has since been extended to account for the effects of debt highlighted by Minsky[Min82, Min86, Kee11, GH15][82].

10.3.1 BULK QUASI-EQUILIBRIUM AVERAGES

A Goodwin model is a mean field theory. One looks for 'monolithic' characterizations of the economy: i.e. single variables that summarize the totality of microscopic behaviours for all agents on average. This approach has worked well for the average properties of cities over decades, where broad universalities have been discovered across a range of scales[BLH+07, Bet13]. It allows us to explore the generic aspects of the major flows. The disadvantage is that the result says nothing about the experiences of any single agent, at any scale, within the model. Society is treated at the level of a ball on a string[83]. Alternative approaches to society modelling have been considered in [Gal12].

10.3.2 MODEL OF A SINGLE CURRENCY ECONOMY

The parameters in a model are generally chosen to represent a single imaginary nation state, with a closed economy. Modelling a world of interacting nation states is possible in principle, but would shift the focus towards even less relevant variables. A loss of detail is a limitation of a mean field theory, but universal scaling laws may mean that such details are not important to the universal characteristics of the model on average. Some evidence of the universality of basic phenomena exhibited by the models comes from [GH15, GH16]. It is plausible that this approach might best be replaced by agent-based modelling, as has been used for cellular automata. Agent-based models are more complex, but more realistic, and offer great potential to gain a detailed insight into the effects on very specific agents within a society.

Minsky's has pointed out that, as a dynamical feedback network, the economy is basically unstable. Goodwin models seek to model those instabilities[84].

10.3.3 QUANTITIES

A challenge for macro-level modelling, in which one deals with a highly aggregated picture through composite variables, lies in the ambiguity of its interpretation. Common terms like 'capital', 'investment', 'consumption', 'employment', etc., are used as though they have unambiguous meanings, but different authors define these aggregate quantities differently. The relationship between labour and production is also simplistic in the age of increasing automation and 'gig' economy.

A Promise Theory approach, in which agents are referred to by functional role rather than '(un)employed', 'investor', etc. may replace this in a multi-role population. Some nation states may actually be vulnerable to an absence of certain roles, certain industries, etc. It is well known that the culture of investment is very different in different geographical regions, leading to international cooperation and investment across borders. Such matters can only be simulated through stand-in variables in an aggregate approach, so there is great scope for improvement in these semantics using agents.

Table 10.2 shows a helpful balance sheet of monetary exchanges used in[GH15] and [GH16] Grasselli and Huu's version of a Goodwin model. This version contains extensions to explore the role of savings, inflation, and debt. We pick their version of the models for their neat presentation.

10.3.4 THE NORMAL MODES

Authors characterize the normal modes of these oscillatory equations by various means [BYLMKG17, GH15, GH16], both in terms of equilibrium attractors, and monetary

	HOUSEHOLDS	FIRMS		BANKS	SUM
Balance Sheet					
Capital stock		$+pK$			$+pK$
Inventory		$+cV$			$+cV$
Deposits	$+M$ also $+D$			$-M$	also $-D$ 0
Loans		$-D$ also $-L$		$+D$ also $+L$	0
Sum (net worth)	X_h	X_f		X_b	X
Transactions		current	capital		
Consumption	$-pC_h$	$+pC$		$-pC_b$	0
Capital Investment		$+pI_k$	$-pI_k$		0
Change in Inventory		$+cV$	$-cV$		0
Accounting memo [GDP]		$[Y_n]$			
Wages	$+W$	$-W$			0
Depreciation		$-p\delta K$	$+p\delta K$		0
Interest on deposits	$+r_m M$			$-r_m M$	0
Interest on loans		$-rD$		$+rD$	0
Profits		$-\Pi$	$+\Pi$		0
Financial Balances	S_h	0	$S_f - p(I_k - \delta K) - cV$	S_b	0
Flow of Funds					
Change in Capital Stock		$+p(I_k - \delta K)$			$+p(I_k - \delta K)$
Change in Inventory		$+cV$			$+cV$
Change in Deposits	$+\dot{M}$			$-\dot{M}$	0
Change in Loans		$-\dot{D}$		$+\dot{D}$	0
Column sum	S_h	S_f		S_b	$pI_k + c\dot{V}$
Change in net worth	$\dot{X}_h = S_h$	$\dot{X}_f = S_f + \dot{p}K + \dot{c}V$		$\dot{X}_b = S_b$	$\dot{X}$

Table 10.2: Balance sheet and transactions flows, cited directly from [GH16]. Note that the authors changed notation for deposits and loans between [GH15] and [GH16], leading to some unfortunate confusion over use of 'D'.

exchange flows. Flow characterizations focus on the two main competing loops of the Goodwin model and whether one dominates over the other. When investment in new capital dominates, wages must fall and consumption may falter, leading to a deflationary spiral of falling demand. On the other hand, when investment is insufficient, there is insufficient capital for expansion, and there may be insufficient production to meet demand, leading to higher prices and inflation.

- Good equilibria in which there is finite debt, positive wages and positive employment.

- Bad equilibria, in which there is infinite debt, wages fall to zero, and employment the same.

The assumptions are that there is a single uniform market, with no explicit dependencies, only a single level of price and wage inflation, and all companies are represented by this singular average behaviour. Money cannot be created or destroyed, so the models do not model the changes in money supply. The benefits of the models are the relative simplicity, controllability of parameters to explore the solution space, and conduciveness to differential modelling. The downsides are the over simplification of the key variables, and the assumption that everything produced gets spent or reinvested like a Kirchoff electric circuit. The models one describes apply to everyone and yet to no one, and their idealization places their effects on a timescale of years that is too long to be of relevance to individuals in society. For an individual, one can go from riches to rags overnight.

Grasselli and Huu[GH15, GH16] have generalized the models to include a number of other parameters such as inventory accumulation and market speculation, and have studied which assumptions are stable to changes of assumption[85], as well at our own notation. They introduce some additional variables, and have a reader-friendly presentation of their model, with attention to mathematical correctness. They consider derivatives $\dot{M}, \dot{D}, \dot{V}, \dot{a}, \dot{N}, \dot{p}$, i.e. the growth rates of deposits, debt, inventory, productive output, population, and price levels.

10.3.5   MODEL TIMESCALES

To understand the limits of validity for these equations, we need to understand to what extend these time derivatives represent actual changes in an economy.

- Timescale of transactions (B2C) (days, months).

- Timescale of accounting (B2B) (days, months).

- Timescale of responses and defaults (days, months, years).

Detailed balance conditions that would accommodate smooth continuity of a differential equation may be achieved in one of two ways: by parallel (spacelike) ensemble aggregation, or by serial (timelike) ensemble stability. The latter case takes longer to achieve, and the former case requires a larger system to absorb fluctuations. Thus the differential macro-models aim their sights at a world of large long-lived corporations.

Grasselli and Huu consider price variations and inventories on top of Keen's model and show that not only debt crises can destabilize economic flows, but also voluntary or imposed hoarding, such as by lack of demand, may lead to collapse of wage share and unemployment. So a recession may also be caused by too many unwanted commodities.

They also effectively show that a lack of money supply, in the face of rising prices can lead to falling employment with or without a debt crisis.

10.3.6   LIMITS OF DIFFERENTIAL MODELS

What could be missing from a differential equation model? By definition, a differential model cannot properly represent catastrophes[Gil81] and discontinuous changes in a well-behaved manner. However, all failure modes in a network are sudden and catastrophic. For example, the failure of one agent causes a cascade of dependent failures[86].

Differential equations are based on the assumption of sampling the system in small increments Δt. To some extent, the relative sampling rates are adjustable by scale coupling constants in the equations. However, there is an underlying assumption of smooth continuity, which obvious empirical experience tells us cannot be a good guide to anyone's experience of the economy.

Discontinuities abound, whether by sudden changes in interest rates, price changes, the start of new businesses, and the death of others, accidents, weather conditions like hurricane events or cold spells that increase heating costs, or personal tragedies, winning the lottery, etc. None of these effects can easily be argued within the framework of a differential Goodwin model. We are limited to regimes in which we can argue:

$$\frac{\Delta t}{T} \rightarrow 0 \tag{10.5}$$

Implying either:

$$T \rightarrow \infty \tag{10.6}$$
$$\Delta t \rightarrow 0 \tag{10.7}$$

Since cash flows are usually computed say monthly or quarterly basis for most businesses, then this is the sampling rate of a trend that could be captured by an aggregate differential model. So, whether appealing to the Nyquist frequency of the sampling rate, or to general physical handwaving, one would expect $T \gg \Delta t$, say by an order of magnitude. This indicates that any changes represented by a dt or a d/dt would be on the timescale of years, not shorter. Thus, a differential model could never capture the sudden catastrophe of crack propagation in a metal or the sudden failure of the banking system by end of month bankruptcy[87].

10.4 A PROMISE APPROACH

It remains for future investigation how to build a full promise-compatible model of scaled economic activity. A differential formulation based on network-wide aggregates, accumulated over an implicit timescale, has a number of weaknesses. It idealizes many of the measures and assumes that there will be no accumulation, buffering, or memory function to monetary exchanges. All these features point to an implicit timescale that is much longer than individual payments. Promise Theory would allow us to understand the scales of any network of agents, and thus offers a bridge between views based on sudden interventions, like policy changes, environmental circumstances, and the slow, broad aggregate measures that refer to the mythical 'economy' we hear about on the news.

The microstructure of any economy must have the form of a network, formed by voluntary relationships of varying durations, each cemented by the binding of various promises. Some influences are non-voluntary, such as the 'exogenous' effects of weather and resource starvation. These promises have a wide range of types. Promise theory describes the scaling of agent-like structures, in hierarchies and flat ensembles[Bur14, Bur15, Bur16a]. The economy may then be seen as a superposition of multiple sparse networks labelled by these types, that interconnect through dependencies. Such conditional dependencies of promises are the link that can transmute one kind of promise into another, forming a broadly percolating network:

$$A \xrightarrow{X|Y_1,Y_2} A'$$ (10.8)

e.g. if I am promised Y_1 and Y_2, I can promise X. The linkage through dependencies is both the source of all component assembly, and commoditization of component construction, to ascend the evolutionary complexity ladder by combination and recombination. It is the also the source of its fragility.

Dependencies are the mechanism that causes errors, failures, and faults to propagate. They imply strong coupling, with brittle transmission of failures. Failed promises propagate at the rate at which the promise binding is relied upon (i.e. the source information is sampled, in information language). Since each agent may keep an inventory (a memory buffer) of what it promises (or what it relies on), agents can be self-sufficient or autonomous for a finite duration.

Example 71. *If a garage stops promising fuel, a car can still drive for many miles before this dependency on fuel affects its ability to promise transportation. Conversely, electric lighting has no buffer for electricity, so its dependence on its power source is strong and immediate, leading to instantaneous fault transmission.*

This tells us that short timescales are a critical part of an economic model, just as debts and repayment schedules incur artificial costs through interest, delays might lead to degradation (e.g. food in a freezer is ruined after a few hours if it loses power). Any default on a time limit is a sudden event that can begin the propagation of a fatal crack in an unstable system.

Example 72. *There is a connection to smart spaces, as defined in [Bur16a], where we have access to economic scaling data, thanks to the ingenious research of West and Bettencourt and collaborators. Such data allow one to see how aggregation affects certain relationships in the presence of dependencies, and gain some insight into the percolation thresholds of promises. This, in turn, makes a connection to more general network science[New03].*

Promise theory allows us to make the connection between the mean field quantities and their underlying elementary sources. Contrasting with a differential continuum approximation, a promise network will be a fundamentally discrete transactional system, that could be used as a blueprint for a cellular automaton or agent-based model.

10.5 SUMMARY

Macroeconomics appear to decouple from the story of money on a small scale. This is partly expected, but it is a weakness that makes its relevance dubious. We can also try to build up a macroscopic picture as a proper exercise in scaling of semantics and interactions—from the bottom up—using Promise Theory. This would clarify the status of assumptions more clearly, and retain a path to explaining how timescales and influences on one scale might influence others. There are many challenges for future researchers in this manifesto.

CHAPTER 11

CRYPTOCURRENCIES

The electronic payment revolution is well underway—especially in China, which currently leads the world in a range of technologies. Over the past decade, cryptocurrencies and the so-called blockchain have dominated the discussion of electronic payment, probably unfairly. The use of Promise Theory to describe these is extremely natural, so we permit ourselves to use this as a case study of the promise language in this chapter.

Blockchain is a class of software technologies, used for building cryptographically-verifiable or tamper-proof 'ledgers' (serial transaction databases), usually built on top of redundant peer to peer (P2P) distributed consensus networks[BSAB$^+$17, Hub18]. As such, blockchain is an example of a system that works by *voluntary cooperation* between autonomous agents, which makes it a natural candidate for description in terms of Promise Theory[Bur05a]. Blockchain was proposed together with the BitCoin cryptocurrency[Nak08][88]. In 2018, the BBC reported some 1300 cryptocurrencies [BBC18]. Additionally, there are applications of blockchain from logistics companies to online universities.

There are multiple blockchain designs, but all of them promise properties such as immutability and redundancy of data, voluntarily cloned across member nodes of the peer networks. They are not unique in making these promises.

11.1 OVERVIEW

Blockchains use digital signing (also called hashing), along with public-private key (PP-keys) cryptography[Bis02], to build a chain of clear-text evidence about past information transactions—in such a way that their integrity can be verified in order to prevent or expose tampering[89] . The key promise of a blockchain database is thus *immutability of data with public transparency.* This has obvious applications for the archiving of financial transactions, deeds of ownership, and all forms of digital governance, where disputes may arise or records are at risk of tampering. For news and historical records, it would also be useful to prevent future parties from rewriting history in their own image. On the other hand, irrefutable data with personal implications and consequences may prove harmful to individuals, e.g. witness protection programmes would become untenable if public identity records were ever committed to blockchains.

Any data could be stored in a blockchain, though proof-of-identity applications such as digital currencies, ownable assets, passport records, and logistical hand-overs, are natural candidates for open transactional records. The first national election in Sierra Leone was recently held via a blockchain system[Big18], there are crowd-funding platforms[Coi], and financing of small loans is now being offered[BBV18]. The deeper question will remain concerning the longevity of data, and whether open access to these data is truly in the public interest[90].

A blockchain database can be constructed on a single computer, but the real power of the technique lies in it being replicated many times, building immutability by engineering massive redundancy against the possibility of brute force attacks against the crypto-graphic integrity. This strategy makes the likelihood of tampering vanishingly small. A replicated blockchain (what we shall call a shared blockchain consensus network) implicates a 'significant' network of communicating agents, which provide redundancy and confirmation by voting. Some notable features of a such a database are:

- A dependency chain of data with cumulative interior integrity checks, within the database.

- A network of peers who promise to keep precise copies of the entire database.

- A consensus mechanism for agreeing on the correctness of a single version of the data, during dynamic updates.

- Incentives for keeping these promises, applicable to so-called untrusted agents.

There might be any number of variations in how these promises are kept, but each application requires distributed members of the network to keep a relatively uniform set of core promises. Some of the promises rely on an assumption about the distributional entropy of peer makeup and location to provide sufficient variety to avoid corruption of the voting mechanisms (just as in any election process).

A key issue will be why a collection of agents would agree to such a system. Any distributed data structure is vulnerable to difference of intent. A consensus system is designed to have singular intent.

11.1.1 AGENTS

In the context of Promise Theory, agents are the irreducible actors, which embody intent, at a particular scale. Agents may be humans, processes, businesses, etc. They are always proxies for human intent. We normally refer to them, in their roles, with respect to information transactions, which are promises to be recorded in the ledger. S is a sender of what is intended (money, information, etc), and R is a receiver (or intended benefactor)[91].

Definition 104 (Node N_ℓ). *A computational instance, running the blockchain software, labelled by its identity location ℓ, that can keep the promises of the blockchain network.*

Nodes form peers in a network of actors, which are executors of the blockchain system. One copy of blockchain data is kept by each full node; some partial nodes may only observe parts of the chain without the ability to add to it. They can promise to take on roles, such as master, slave, miner, etc, that equate to the keeping of certain sets of promises. The users of the blockchain system are clients.

Definition 105 (Client user). *A human agent, who imposes a transaction into the blockchain, or who is the recipient of a transaction. Clients thus have the roles of sender S or receiver R of transactions.*

A client promises an identity via a public-private key pair. A single human agent may have multiple key pairs[92].

The purpose of the blockchain is to record transactions for clients.

Definition 106 (Transaction τ). *The imposition I_τ of a request, by a sender client S to a blockchain node agent N_ℓ, to commit information $\tau(S, R, \ldots)$ to the blockchain.*

Transaction requests are *impositions* of promise proposals into the blockchain system, in promise theory language. If accepted, those imposed requests become promises π_τ by blockchain nodes to record the transaction proposal at some later time within a data block. Data blocks, thus form the elementary units of a blockchain, are also agents.

Definition 107 (Blocks β_a). *Data blocks are agents that promise to aggregate, store, and validate transactions. Data blocks also promise their own integrity or tamper-proofing, using cryptographic hashing in a number of ways.*

Within β_a, transaction encodings promise to use hashes of earlier transactions to form a Merkle tree, providing cumulative verification of integrity. This is a promise of interior integrity. Further, each block β_a promises to be a candidate successor of an earlier established block β_{a-1} ($a > 0$), thus forming a chain of blocks. Finally, the block promises that all transactions, linkage, and other padding values all are verifiably unaltered. This constitutes a promise of exterior validation.

This final tamper-proofing comes in a number of varieties. Poof of work, and proof of stake are two methods currently discussed. The original idea was that whatever scheme is used, it should levy a huge cost on a potential tamperer. This philosophy is changing however, as the costs of processing such penalties (e.g. in terms of energy expenditure) begins to dominate the discussion[BSAB+17].

Blocks do not and cannot promise their provenance, i.e. that they originate from a particular location, but through their design they can promise to require a unique investment in truth, using these proof of work or stake schemes.

Definition 108 (Blockchain C_ℓ). *A collection of data blocks, or structures, which are linearly ordered and successively dependent by cryptographic validation, thus forming a dependency chain.*

Blockchains have verification properties that make them difficult to alter without authorized cryptographic credentials. They therefore belong to a set of encryption verification methods. Blockchains are useful as shared data structures, and this is the application that prompted their origin in [Nak08].

Definition 109 (Shared blockchain consensus network **N**). *A collection of nodes N_ℓ, which form a peer network such that each agent keeps an exact replica of candidate blockchains, until a single chain is chosen by consensus, based on a selection criterion.*

Each specific instance of blockchain technology may provide its own rules about how agents add and delete blocks. To form of shared blockchain consensus network (or what is colloquially referred to simple as a 'blockchain technology'), all agents must do so in such a way that all the agents eventually promise to agree about a single linearized collection of data blocks. This is what we call the distributed ledger.

Definition 110 (Blockchain Distributed Ledger). *A ledger stored and published in the medium of a blockchain consensus network.*

11.1.2 PEER TO PEER ARCHITECTURE

A key component to blockchains is the distributed voting mechanism, which is used both to select the official record, and secure it against loss. This is a form of distributed consensus network, more usually implemented by very different technologies.

A *consensus network*, or *quorum* is a network of nodes in which every node promises to hold the same information at all observable times (see figure 11.1), and all agents agree on what 'the same information' means. In the case of blockchain, the nodes agree

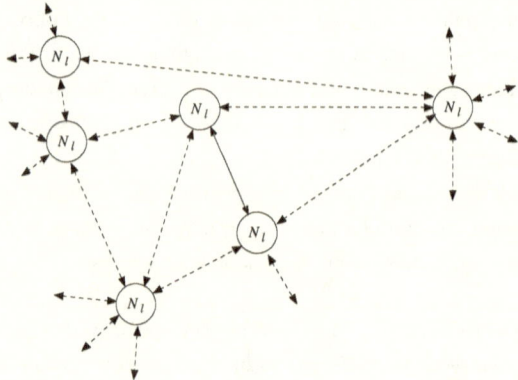

Figure 11.1: A peer to peer network $\mathbf{N} = \{N_\ell\}$, in which each node (or instance of blockchain software, running on a computer) communicates with its nearest neighbours. This is also the basic picture of agents in promise theory; however, in that case, directed links would refer to specific promises rather than a generic ability to communicate.

about the structure and content of a single data structure called the blockchain, which they are (a priori) free to read and append.

In an open blockchain, nodes promise to forward or broadcast all blocks β_a and transactions τ to all agents of their peer network:

$$\pi_{\text{consensus}} = \left\{ \begin{array}{c} \{N_i\} \xrightarrow{\pm\beta,\tau} \{N\} \\ \mathbf{N} \xrightarrow{\pm\beta,\tau} \mathbf{N} \end{array} \right. . \tag{11.1}$$

though this is too vague to deliver much insight. In practice, nodes make many promises, e.g. the software promises to listen on port 8333, to receive messages, and to service imposed transactions by remote parties, etc. Nodes must also promise sufficient memory to hold a complete copy of the datastructure C, else they cannot be allowed to represent it faithfully.

$$N_\ell \xrightarrow{+(\text{memory}_\ell \gg C_\ell)} N_{\ell' \neq \ell}, \quad \forall \ell \tag{11.2}$$

These promises are trusted requirements for the blockchain. Thus nodes, and the software they run, have the status of trusted agents[93][BB06b].

The blockchain datastructure is replicated on every node that keeps the full set of promises. It remains somewhat unclear whether validating all blocks from the beginning is a mandatory requirement of blockchains, or whether a chain could be picked up at any location to verify subsequent promises. It is composed of blocks β_a, which resides within a local copy of the globally agreed chain. Each local chain C_ℓ, on the node N_ℓ is a connected collection of all approved blocks β_a:

$$C_\ell = (\beta_a, \pi_{\text{chain}}). \tag{11.3}$$

Approved blocks are not independent, but form a chain (i.e. a recursively defined pipeline, which is used for validation), by promising to base their content on the content of prior blocks. Each block β_a promises a hash h_a of the preceding blocks.

$$\pi_{\text{chain}} = \begin{cases} \beta_a \xrightarrow{+h_a(h_{a-1})} \mathbf{N}, \\ \beta_{a+1} \xrightarrow{-h_a} \beta_a, \qquad a > 0 \\ \beta_{a+1} \xrightarrow{+h_{a+1}(h_a)} \mathbf{N}, \end{cases} \tag{11.4}$$

making the blockchain promise a kind of totally ordered peer network for the block data, at the level of the datastructure. In a closed blockchain, only privileged agents would be able to assess the state of the chains. In modern developments, the use of chains is being extended to the use of generalized tree structures (Directed Acyclic Graphs, or DAGs).

Blocks make structural promises so that any node can verify the validity of a block quickly. Conversely, it takes a costly effort by a highly committed node to produce this validation, making it difficult to create valid blocks that might be used to tamper with existing data. There is an increasing flora of varieties for how this works, with different optimizations.

Example 73. *In the original BitCoin ledger, for instance, this promise is called 'proof of work': updating the datastructure is necessarily a relatively slow operation, because achieving a consistent certainty and agreement (consensus) on the content of the data, distributed amongst so many nodes, is a slow process. In the BitCoin case again, each mined block promises its own hashed verification value by adjusting a 'nonce' dummy field; this amounts to a self-consistency promise that is quite calculate, and thus blocks cannot be altered without recalculating this validation.*

Self-consistency promises, in turn, make it very difficult to alter any information that has already been accepted into a blockchain. So the blockchains approximate (very well) an immutable data structure that is replicated with a high level of redundancy.

11.2 DISTRIBUTED CONSENSUS SYSTEMS

A 'shared blockchain consensus network' belongs to a class of technologies called distributed consensus systems, whose purpose is to distribute precise replicas of a database to redundant locations. In other words, the goal is to engineering a common knowledge about data across a realm composed of multiple independent agents[94]. Well

known consensus schemes of this kind include View-stamped Replication[OL88],
Paxos[Lam78, Lam01, GL06], Zookeeper[JRS11], and Raft[OO14]; all of the foregoing
make use of an authoritative master node, through which all transactions must pass.
Blockchain works differently to most such scheme, however, so we review the concepts
briefly here.

Arriving at a consensus is basically the same problem as calibrating data to an
agreed standard. First one has to define the standard, then ensure that all agents follow
it. Calibration thus implies that a single standard of behaviour is upheld with respect to
processing of data too. The most obvious way to do this is to send all data to the same
place, i.e. serialize data through a single service gateway with constant semantics. This
simulates a simple causal determinism, and avoids unwanted effects of relativity, where
different messages arrive at different times due to the finite speed at which messages
can travel. It helps to select a single spatial location for this point of calibration (called
a master node) , else one may have to deal with these signal delays, complicating the
selection of a standard[95]. However, this is not the approach used by most blockchains.

Distributed consensus systems thus come is two main varieties: single point of
definition and equilibration by averaging, making use of the two fundamental aspects of
a system, dynamics and semantics:

- **Local dynamical ordering (local spacetime causality).** Most consensus schemes
  involve the election of a master node from a privileged set of custodians serving as
  calibration service. As data change, only a single location needs to be updated, i.e.
  a single point of change funnels all change requests, which it then copies exactly
  to nodes that keep slave copies (assuming no sharding of data[96]). A master system
  relies on a predefined quorum on the authoritative data, by identifying authority
  with the identity of a master source for all subsequent transactions[97].

- **Semantic ordering (criterion ranking).**

  Where one cannot rely on correct order of arrival, or even a fixed location, we have
  to order arrivals based on semantic attributes. This may require a computation
  for the selection process, applied to every transaction. All agents receive data
  independently, from different sources, and possibly in different orders. There is a
  'race' for transactions to arrive in a particular order, and a semantic selection has
  to be used settle any contention about ordering, e.g. labelling the data in order,
  selection of the best or longest chains, etc. These require agents to remember
  prior data, and vote on a post-defined quorum for which version of ordering is
  authoritative. Blockchain takes this approach, allowing it to reach consensus by
  broadcasting and averaging the state of all neighbours[98]. Distributed averaging is
  a slow process, because it may be up to $O(N^2)$ in the length N of the queue.

In both cases, a voting process is used to build a quorum (before or after accepting transactions) understanding of which version of data is the authoritative one[99].

Example 74. *The integrity of the ordering is protected by a cost of entry for agents broadcasting blocks, which is paid for in thermodynamic work. BitCoins daemon can quickly separate valid from invalid states, once they are in the stream, because blocks cost a lot to prepare (about 10 minutes of CPU time). It is thus not 'possible' (likely in practice) to spoof fake information into the broadcast stream, without a large effort. This is not a foolproof discriminator of trusted membership, but it is considered effective in practice for now.*

Distributed data replica systems are typically discussed according to two semantic standards: so-called ACID (Atomicity, Consistency, Isolation, and Durability) or BASE (Basically Available[HW90], Soft state, Eventual consistency) characteristics. The former implies that data are immediately consistent; the latter implies that data may become consistent after some equilibration time has elapsed, which assumes that intentionally equivalent data are held immutable once submitted. All systems have eventual consistency in practice. The primary issue is whether or not observers are allowed to see the states of intermediate consistency or not. This is controlled by the software service's promises to its clients.

Since equilibration time is crucial to the promise of observational invariance (not just covariance), data propagation speed (or latency) is a key factor, as is the definition of the event horizon for data. Transport congestion and observational security both play key roles in this.

11.2.1 KEY PROMISES FOR CONSENSUS AND BLOCKCHAIN

Ledgers may make a number of promises about their particular characteristics[100]. These are not always clearly stated. Reference [ABSB+17] is a helpful paper in that it makes very clear statements about its claims. Typical issues include:

- *Open or Closed*: The policy for the ledger. If any agent can join the system, and is possibly rewarded for participation, the system may be called 'open'. If special credentials are needed to participate, then the system may be called 'closed'. Closed systems are more natural in cases where personal information or privacy concerns are paramount.

- *Atomic transactions*: All databases promise some kind of atomicity. A transaction is an atomic computation, (not necessarily just an atomic nugget of information). A transactional system should make clear promises about how the execution of contracts, or promise-keeping events, will be encapsulated, and what will be observable by whom.

- *Transparency*: Various levels of transparency may be promised. The level of transparency of an information system refers to the extent to which source code, transactions, and states of computation can be inspected by all agents (or just some).

  In any decentralized system, we have to assure the security not only of intent but also assessment. Promises (operations) and assessments (monitoring) can both be distorted by incentives, faults, and errors.

- *Integrity and Non-Repudiation*: A promise of integrity refers to the idea that transactions will be executed exactly once, accurately, with a total ordering defined across the cluster. Non-repudiation is a promise claiming tamper-proofing, or at least detectability of tampering.

- *Encapsulation (smart contracts)*: In the case of transactional computations, different architectures may promise difference isolation mechanisms. Transactions should be clearly separated, without unintended leakage between separate intentions.

- *Auditability and regulation friendliness*: Another promise of transparency refers to auditability. No coded software can be trusted completely to encapsulate something as complicated as systems, which today are governed by complex legal frameworks. Complying with complex societal intentions and guidelines is non-trivial, so there needs to be room for assessment.

  Blockchains have been called 'trustless' (sic), but this does not mean that they can replace trusted human institutions like the law.

- *Proof of X*: e.g. proof of work, proof of stake, etc. Each blockchain promises to validate participants somehow.

In addition to these technical promises, there is any number of proposed benefits claimed. There is a tendency towards over-promising what a mere data structure can achieve in software circles, namely that software alone can replace the malfunctioning and even possibly corrupt institutions of society. We shall comment on some of these below.

11.2.2   COMPARING CENTRALIZED AND DECENTRALIZED LEDGERS

What promises might a blockchain keep that a centralized master-slave system (with backups and redundancy) might not keep? Both the operation and the storage of data within master and non-master consensus systems are quite different. We can use promises to compare and contrast these approaches. Figures 11.2 and 11.3 are simplified, schematic promise graphs that attempt to illustrate the difference in information architectures between the two cases.

- A centralized redundant ledger, built on a master-slave principle, accepts transactions (unconditionally) τ_{ij} submitted only to the master node M by clients C_i, and these are broadcast serially to slave nodes.

$$C_i \xrightarrow{\ +\tau_{ij}\ } \blacksquare \qquad M \tag{11.5}$$

$$M \xrightarrow{\ -\tau_* \ |\ \text{quorum}(M,S_*)\ } C_i \tag{11.6}$$

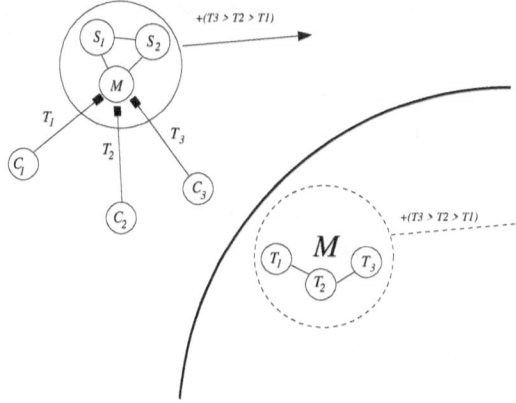

Figure 11.2: A centralized redundant ledger built on a master-slave principle. Transactions are submitted only to the master node, and these are broadcast serially to slave nodes. Transactions are stored directly in nodes, and the master node's version of reality is the authoritative one. A quorum has to exist *before* acceptance of transactions by the master, to avoid accepting an inconsistent state. Thus equilibration takes place in a typically small set of master-slave nodes. On the right, an exploded view showing the arrangement of transaction agents within each of the master nodes, ordered for each client by First Come First Served.

The master node promise replication to its slaves:

$$M \xrightarrow{\ +\tau_*\ } S_* \tag{11.7}$$

$$S_* \xrightarrow{\ -\tau*\ } M \tag{11.8}$$

Transactions are stored directly in the master node, and the master node's version of reality is the authoritative one.

An election quorum, about selection of an authoritative master, routing transactions through the master, and replication of information has to exist *before* acceptance of transactions by the elected master, to avoid accepting an inconsistent state. Thus equilibration takes place in a typically small set of master-slave nodes. On the right of the figure, an exploded view showing the arrangement of transaction agents within each of the master nodes, ordered for each client by First Come First Served.

A master node is a single point of failure, but if it should fail, a new election can take place to replace it, taking over where the master left off, uninterrupted. So, with sufficient redundancy centralization is not a specific risk.

Thus the causal sequence is: Global quorum for all transactions → Feed into master node→ Replicate to redundant cluster

- A decentralized ledger, based on blockchain, works without the tight coordination of a master-slave cluster. A chain is a conditionally dependent structure, with the its inherent fragility[101].

In a blockchain peer network, transactions are submitted to any or all the nodes, and they are broadcast to all nodes. The nodes select the transactions they like the look of, possibly swayed by incentives, and blocks are mined by racing nodes. The winner extends the blockchain, and broadcasts the new block change. Other nodes voluntarily accept the blocks and promise to break ties impartially using a longest chain criterion to select an authoritative version. Transactions are thus aggregative in blocks, which or stored as a chain within nodes. Each node's blockchain promises an authorized (and eventually common knowledge) total ordering of transactions, by the agreed acceptance of a longest chain. Each block is a single point of failure for the chain, but copies of missing or broken blocks can be obtained from other peers to counter this threat.

A quorum is formed (eventually) *after* transactions have been accepted into blocks on nodes, and competition amongst all the nodes establishes an equilibration of state across a much larger set of authoritative nodes than for a typical master-slave system. Inside blocks, transactions are ordered for each client by First Accepted First Served. The order is not First Come First Served, because service order may be distorted by differing path length latencies and incentive fees.

Replicate transactions to peer network → Mine into blocks → Find quorum per block.

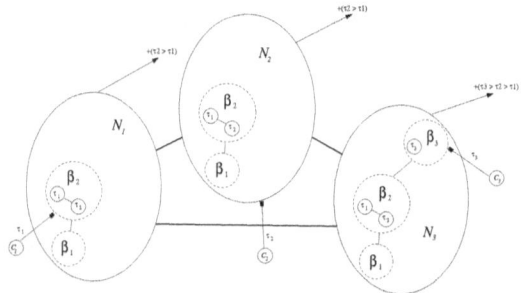

Figure 11.3: A decentralized redundant ledger built on a blockchain principle. Transactions are submitted to any and all nodes, and are broadcast to all nodes. Blocks are mined by racing nodes, and the winner extends the blockchain, and broadcasts the change. Other nodes promise to adopt the longest blockchain they can see as the official one. Transactions are thus stored in blocks, which or stored as a chain in nodes. Each node's blockchain promises an authorized (and eventually common knowledge) total ordering of transactions, by the agreed acceptance of a longest chain. A quorum is formed (eventually) *after* transactions have been accepted into blocks on nodes, and competition amongst all the nodes establishes an equilibration of state across a much larger set of authoritative nodes than for a typical master-slave system. Inside blocks, transactions are ordered for each client by First Accepted First Served. Order may be distorted by differing incentive fees.

11.3 TRUST AND BLOCKCHAIN

Trust is the level of belief an agent has in predicting the outcome of a promise, before any assessment or validation is made[BB06b]. If every action or behaviour is verified, this is an indication of zero trust. If no actions are verified, this is an indication of complete trust (which may be normalized to a value of 1).

Whether agents can be considered trustworthy or not is a matter of much interest and debate in information systems. A common attitude is that trust is a sign of vulnerability and that verification is a necessary and sufficient cure for this weakness (one might call this trust in trust)[102]. In fact, trust is something of a phantom, which can never truly be eliminated. So, to evaluate trust in blockchain, we need to be clear about what is actually promised amongst these various claims.

Blockchain currencies are currently global. However, one might imagine political interests may eventually lead to conflicts over what transactions one would be willing to accept when currency is independent of political union. Blockchain currency is already banned in a number of countries.

11.4   TRUST IN BLOCKCHAIN

Blockchain technology has been called trustless (sic) or trust-free, in relation to its use of agent consensus, block validation, open access to new participants in the network, and the incorporation of data integrity validation by all participants as part of the peer protocols. A blockchain network is self-validating in this limited sense. Nevertheless, the idea that no trust is needed (which paradoxically implies that one should trust it completely) is a bold and quite misleading assertion. A more pertinent question is to ask which promises we are trusting.

Trust in services is usually approached by appealing to the integrity of a single master agent for the authorized handling of key promises according to a common standard. It is not common to include reliability in trustworthiness, only semantic correctness. This is more problematic in a distributed system, because a correct response is more sensitive to whether data arrive at a particular location. A singular agent who performs a service at an authorized location, is often referred to as a Trusted Third Party (TTP). Banks play the role of this trusted intermediary for transactions of regular bank money (either as a direct electronic service point, or as the validator of authorized coins and notes). A single TTP calibrator allows clients to believe in the *consistency* of the agent, with respect to all clients.

If one does not believe in the integrity of a single agent, then multiple agent oversight (exterior regulation) is possible. With multiple independent agents cross checking, one can seek a quorum or consensus in the assessment of promises[103]. In blockchain, promises are cumulative and thus trust is built cumulatively. Older blocks are more trusted than newer blocks, since the leading edge of the chain may still be disputed.

There is some overlap between these two apparent positions, if we take dependencies into account. Most services rely on a standard kind of software, even if there are many service points distributed around in space. Agents are usually not completely independent; they typically run the same software, so the software itself becomes the single point of trust in many cases. Following the 'collapse of trust' in banking institutions after the 2008 financial crisis, software communities therefore proposed that exchanging trust in a banking elite for trust in a software elite was a change worth exploring.

By focusing on a 'single point of trust' as a 'single source of truth' (SST), we form a calibrated standard whose invariance (in terms of promises kept) becomes the same issue as the invariance of the agent itself. A single point removes the issue of horizontal (spacelike) variability within a system, but it does nothing to eliminate any longitudinal (timelike) variability.

A master node may promise to be a single source of truth, but its promise depends

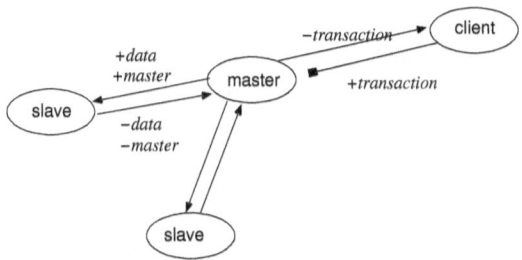

Figure 11.4: A master-slave consensus network. When a node is elected master, that node promises to accept transactions from clients on behalf of the others, and the others promise to subordinate themselves, accepting only authorized updates from the master. In this single-master scheme, transactions are (quickly) serialized by virtue of there being a single queue.

on the sources of factual information fed to it (from distributed sources, see figure 11.2):

$$\text{Source}_1 \xrightarrow{+\text{fact}_1} \text{Master} \tag{11.9}$$

$$\text{Source}_2 \xrightarrow{+\text{fact}_2} \text{Master} \tag{11.10}$$

$$\text{Master} \xrightarrow{+\text{SST}\ |\ \text{fact}_1, \text{fact}_2, \cdots} \text{Clients.} \tag{11.11}$$

The keeping of all these promises is what leads to the semantic and dynamic keeping of the promise by the master node. So we can never really separate trust from reliability. Finally, there is still the possibility of a single source of truth of being a single point of failure. Data corruption can still lead to semantic unreliability of the most calibrated of sources. Redundant systems are only a security against a breach of this trust if clients actually check redundant sources (like good scientists or journalists); it is not enough for there to be multiple copies if only one is used.

The calibration afforded by channelling facts through a single point is not possible in a distributed service, where there is less reason to believe in horizontal consistency of facts between different service points: agent independence and a lack of common causality cast doubt on the consistency. Even if the agent promises are globally invariant with respect to data, they are potentially uncalibrated with respect to behaviour (they may not even be using the same software). Elaborate schemes for data consistency (distributed consensus) may be used to make limited promises about horizontal variability of data values.

11.4.1 VALIDATION OF DATA, AND RISKS OF IMMUTABILITY

One of the notable promises of blockchain is to make transactions immutable. Should the inputs to an immutable ledger be validated somehow? The growing phenomenon of

'fake news' and the shift from a trusted information society to a reputation society means
that immutability could quickly become a liability. Impartiality, as journalists, news
organizations, and scientists once promised, seems harder and harder to trust. Blockchain
is irreversible computing.

Tamper-proofing cannot be promised, because we cannot promise that something
will not happen. What can be promised is the cost incurred by an agent trying to tamper
with data in each system. Blockchain tries to make this cost high, while master-based
systems allow it cheaply.

How do we assess the potential damages caused by undoing information? References
and dependencies on earlier promises can become orphaned if those earlier promises are
rescinded. Cleaning up all subsequent dependencies of a change may be very difficult. It
is usually impossible. However, this can also happen in a blockchain if a user loses their
PPkey, so the risk of loss is present with different profiles in both systems.

A key feature of interest in blockchain is that horizontal variability in blockchain
data is both detectable and resolvable, provided there are sufficient nodes online, and that
one assumes the horizontal consistency in the software promises kept by the nodes.

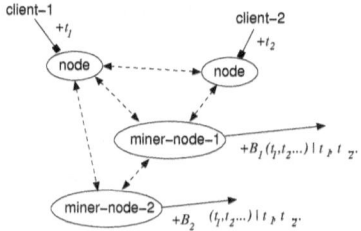

Figure 11.5: A masterless (blockchain) consensus network. Any node can accept transaction
proposals from clients at random. All nodes promise to broadcast these proposals to the network
by gossiping to their neighbours in the P2P network. Nodes that promise the role of 'miner' may
accept the transactions and try to aggregate them into a block. Such blocks can be produced by
any node, and are also broadcast to all nodes, who promise to accept them if and only if it satisfies
criteria authorized by the blockchain software. In the end, only one block from one miner will
be accepted, and there is no guarantee that a given transaction proposal will be processed. In
this masterless scheme, there are multiple queues in competition, and transactions are (slowly)
serialized implicitly by a final selection criterion. Non-determinism implies that users cannot
automatically trust that their transactions will be honoured.

> **Definition 111** (Blockchain's trusted promise). *Blockchain's major claim is that, malicious nodes do not need to be independently trusted because any foul play on their part would be reflected in block data that would fail verification, and thence be overwritten by honest and rational agents, according to the incentives offered within the system; this relies on the availability of a quorum of sufficient size at any time in the future at which validation might be sought. This assumes that clients actually check multiple sources rather than trusting a single node.*

The data consistency is handled by a validation mechanism based on:

- Trusted calibration of node software according to behavioural rules nodes have promised to obey.

- The trusted broadcasting of transaction and block data to all peers.

- A deliberate competition between agents, with trusted cryptographic tie breaker.

- The cryptographic validation of data blocks using a trusted hash function.

- Agreement (consensus or quorum) by a majority of nodes checked, and trust in this quorum.

All these steps assume the consistent keeping of promises encoded into the blockchain software, and the belief in safety in numbers.

In practice, the cumulative nature of blockchain means that the threat of undoing blocks dwindles as they become older and the threat of dominating and overwriting the official blockchain becomes less. In other words, there are grounds to trust the integrity of older blocks more. However, it is not beyond the realm of all speculation that a major attack could be mounted against either a single user or a partition of the network: a gradual attrition of interest in participation, or legal rulings causing nodes to drop out of the blockchain network, followed denial of service attack on the network to rewrite records is not beyond the scope of plausibility. Believing that what goes into a blockchain is immutable could be a serious new error of trust. The counter point is also important. The inability of blockchain to forget anything could be equally problematic, as malicious users abuse the immutability of the medium, like graffiti vandals use indelible markers to despoil public spaces[Cla].

11.4.2 THE ROLE OF TRUSTED THIRD PARTIES

We see that it is quite inaccurate to call a blockchain a *trust-free* system, as is sometimes claimed (trust in a group, a codebase, or even an ideal, is still trust, or acceptance without validation). Blockchain software may not formally need to be sourced from a single origin, but agents expecting to join the network must keep a set of promises that are effectively sourced from a single specification. There is thus an implicit contract involved in using a blockchain. The policing of this contract is built into the software as far as possible[104]. Most users take this software promise on trust. It's 'trust' all the way down.

It is not the avoidance of a single master node (or trusted third party) which makes a network 'trust-free', rather that no single node acts alone or unverified by others, as opposed to being regulated by a possibly (un)Trusted Third Party (TTP). In a master node consensus system, the TTP is the transaction kernel and the software. For monetary matters it is usually a bank or a broker. A monolithic system (monolithic in composition) is easier to associate with a legal entity.

In a blockchain, the Trusted Third Party is a collective formed by a majority of nodes running the trusted software. In all cases an endless chain of validating turtles (all the way down) would be required to completely avoid trust. Clearly, this is not possible. One can only improve the odds of a trusted outcome being favourable, by cumulative experience. At some point one must give up verification and simply trust.

Blockchains promise that the outcomes of transactions recorded within them will be verifiable, no matter their origin, provided all nodes use the approved software version.

A distributed blockchain, based on peer to peer, requires the cooperation of many different parties. In cases where it's implemented as transparent open source software, it has a certain level of verifiability, but ultimately no one can see which software any node is running.

11.5 PERMANENCE OF BLOCKCHAIN

The wide range of promises offered and employed across the space of distributed consensus systems, even those calling themselves 'blockchain' makes predicting their properties hard to generalize. How should anyone choose a system for consensual record keeping? The main selling point for blockchains is often touted as transparency and immutability without trust. Whatever one might think of the semantics, the practical consequences are worth questioning.

Permanence is not a unilaterally positive property. While technologists tend to focus on the hard problem of keeping data, they often forget that systems need to forget information as part of a natural reduction of cost (energy). This is a purely (thermo)dynamical necessity.

There is also a semantic need to forget information. Too much information is the same as none. Permanent defamatory information of a personal nature may cause harm to reputation: to individuals, companies, or families (during someone's lifetime and beyond). When is it ethical to place information in the public sphere? For example,

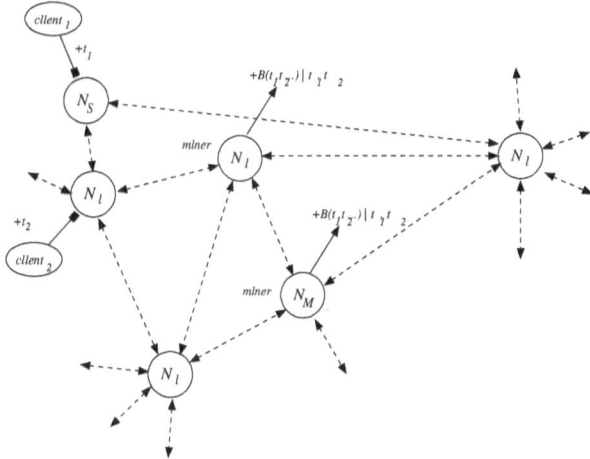

Figure 11.6: In a blockchain network, nodes promise to accept transactions imposed by clients as 'request for transaction', some nodes promise the role of miners. Miners promise to select and aggregate transactions into a new data block, which can be incorporated into the blockchain. Several miners may compete, in a slow race, with the same or different sets of transactions to fill a block which can be incorporated into the blockchain. All nodes promise to broadcast or relay transactions and blocks to their neighbours.

untrustworthy governments have tried to rewrite the history of their countries to eliminate records of unfavourable events, such as the atrocities, weaknesses, or eras of political upheaval. Having an immutable public records 'for humankind' about public history and public news would be a security, where governments and institutions could not be trusted because of political motivations. Witness protection programs would become impossible if identity records were made on public blockchains. Secrecy is still a necessity in some cases to protect parties. To whom do we entrust the decision about this selection and its initial impartiality? There is no technological solution to this matter.

Trust has been shifting away from government institutions (who have been slow and resistant to adopt change) towards to commercial enterprises who offer quick solutions to pressing needs, but who may have fundamental conflicts of interest (e.g. social media companies). Conversely, public institutions may seem more trustworthy as arbiters of economic transactions, based in legal frameworks and the traditions of a long-standing justice system. The introduction of a vigilante system of approvals, or a flash mob, signing off on transactions (which is one way of characterizing a blockchain peer network) would signal a major shift in political power.

11.6   SMART CONTRACTS

Cryptocurrencies have the capacity to extend the semantics of money and price beyond
the simple firewalling of microcurrencies, by replacing the dumb agents of traditional
money with 'smart' reasoning agents, such that the capacity for reasoning follows the
relevant information rather than staying locked up at a central ledger. Centralization
of transactional functions, has played an important role in promising a consistently
calibrated service to users in banking, but it has also limited their potential functionality
as well as their scalability.

Smart contracts (on top of blockchains) are computer programs that keep contractual
promises as an embedded trusted intermediary, which users are supposed to trust by
virtue of trusting the integrity of the blockchain. This is not a full surrogate for truly
mobile code, because the agents of the blockchain and the agents being transacted are
not formally connected. Placing a stashed or cached version of code close to all locations
is not a major innovation. It is only superficially different from a centralized service
repository with a parallelized back end.

A promise viewpoint points towards using the autonomy of agents, and the keeping
of promises by the agents that make them. Why would the code not accompany the
very thing being transacted instead of being stored in a blockchain alongside it? The
immutability argument is one answer to this question, but this brings as many problems
as it solves, particularly with regard to upgrading and rescinding of promises.

Example 75. *Smart contracts have many possible applications. For example, a song
or film posted on the Internet could be distributed, opened, and copied, all by making
a payment which is then returned to its maker. In this way, business projects could live
their own independent lives on the Internet, without the need for human intervention.*

Example 76. *Applications in small scale financing have been discussed[BBV18]. Distributed loaning is nothing new, of course: WeChat and Alipay, in China, offer loans to
through social media accounts, quickly and with excellent terms. The main difference is
the level of transparency behind the scenes, or perhaps interchangeably the level of trust
deemed acceptable in the system.*

Insurance and logistics companies see possibilities widespread and high speed access
to secure services. It remains somewhat controversial just how secure and private
Blockchains could be, given the tension between the need for privacy and transparency.
A deliberate separation of concerns is needed, which could lead to conflicts of interest.

A brief evaluation of smart contracts has a number of dimensions:

- Smart contracts may be executed in permissionless networks which arbitrary
  participants can join (i.e., under Byzantine conditions). This is risky and a

potential security liability.

- Programming languages for smart contracts: Solidity, Serpent, with frameworks for test-driven application development including Truffle (with promises) and Embark.

- Blockchain nodes have meaningful control over the environment in which the transactions execute (e.g. which transactions to accept, transaction ordering, setting of block timestamps, manipulation of call stack, etc.) As with any distributed service, it is not trivial to make assurances on either side as long as there is a non-local behaviour, in spite of an underlying voluntary cooperation.

- The immutability of blockchain records means that there is enormous pressure to get things right first time. For monetary transactions, mistakes can be fixed by voluntarily adjusting payments based on human interactions (human trust must still play a role). However for smart contracts and other services, this is more challenging.

- There is no way to patch a buggy program or smart contract, no matter how much money it has, without reversing the blockchain (which would be an unrealistic task). Users can, naturally, refrain from using promised code, but this opens a new role for trust in the platform.

  Reasoning about the correctness of smart contracts before deployment is now critical, as is designing a safe smart contract system, by some agreed set of standards. One may explicitly design versioning and upgrade capabilities into smart contracts code, since contracts can promise to call each other ([Exc]); but it remains the case that the original byte-code associated with the contract is immutable for all time, with unspecified implications, so the assessment of safety relies on one's trust in the promises of the platform.

In Asia the financial industry still relies, to a large extent, on manual processes and paper forms, often transmitted through fax. This is obviously inefficient, high cost, and involves operational risks and delayed settlements, which increase during peak processing times[Lee], and may lead to insider tampering. For example, if an Asian company invests in assets that are originated by entities in a different jurisdiction, then investors in Asia, no matter where they are based, could seek legal recourse in foreign courts, assuming the blockchain were trusted. This remains untested, however. Remarkably, these questions still remain questions of trust, in spite of the trustless rhetoric. In fact, China as already signalled that it should be illegal to trust foreign blockchains.

The timing and convergence rates of blockchain are still in need of more study. Contracts that are executed, based on records that are still under potential dispute, may become invalid as blockchain forks resolve. Thus, effectively, promise outcomes could be turned into lies by future events. This uncertainty is borne by users of the platform, and must trigger consequences. It suggests that blockchain technologies need to be able to promise users when results are in fact immutable, or when convergence has occurred finally. There remains therefore some question as to how useful complicated distributed systems can be compared to central clearing agency promises.

Finally, a general theme in blockchain is that the naive use of immutability can be as much of a liability as a benefit. Indeed, as time goes on, and the length of blockchains grows, that liability grows with it. A proper design must have the ability to deprecate and garbage collect old data blocks. There is, of course, a sense that the implications of a new technology have not been fully thought through. For blockchain, this issue is more serious due to the immutability.

11.7   SUMMARY

Blockchains, and other distributed ledgers, blend several ideas to fashion a model for achieving distributed consensus around a ledger of transactions that are not regulated by any central 'master' or authority.

Blockchain solves the following problems:

- It offers a form of tamper-proofing, and eternal immutability, for better or for worse.

- It promises to replace an opaque intermediary with a slightly more transparent intermediary, in practice it is no more transparent to ordinary users, who use trusted 'fourth party' applications like wallets and brokers to handle their transactions.

- Blockchain solves the technical challenge of data serialization in a masterless architecture[105].

It does not solve these remaining problems:

- It does not eliminate the need for trust: it shifts trust onto different focal points.

- It does not answer the question of why users should trust a majority vote or quorum rather than a central governing entity. Couldn't this be manipulated to lead to abuse? There is some naivete concerning the objectivity of software automation.

- Proof-of-X schemes all favour those players who have most resources to begin with: most computing power, most memory, most money(!) This tends to speak against the view of blockchains as being the great democratizers of society.

Cryptocurrencies solve the following problems:

- They enable transactions to be made without individual discrimination, though systematic discrimination is still possible.

- A number of technical issues about conservation of currency are handled by the 'double spending protections', but these do not protect users from being duped.

- They protect the 'dilution of the currency' against monetary inflation, which is mainly of interest to institutions rather than individuals. There is no monetary inflation with BitCoin, but this is a purely technical promise with both positive and negative connotations for users.

They do not solve these remaining problems:

- Transaction speed is considerably slower with cryptocurrencies than for most banks, so payment by blockchain may not be suitable for small day to day transactions, without trust in the software to guarantee a transaction time.

- Transactions could still be rejected by the software, once submitted.

- The role of trust in end users becomes more like that in bartering again: cryptocurrency might promise to make an immutable payment, but it doesn't promise that you will get the goods you paid for, nor that a payer will repeat a rejected transaction after receiving the goods. Moreover, the buyer has no recourse to an institution to act as a mediator. The cold logic of the blockchain will be unsympathetic to such broken promises. We have to be aware that this is precisely the reason for human trust relationships in the first place; so, trying to eliminate them will have a cost.

Transparency is the perhaps the most widely used claim for blockchain—but transparency is the opposite of privacy. The initial popularity of blockchain really seemed to show that trust in our civil institutions has been compromised. This is a crisis for future society, but it is not an argument for total transparency.

The impact of sharing information might be positive or negative—indeed, it must be relative to all the parties involved in it. Promise Theory's Law Of Conditional Promises makes a simple but important point: if an agent A promises information I to anyone:

$$A \xrightarrow{+I} * \tag{11.12}$$

and it is used by some party X, who then derives some followup promise F based in I

$$X \xrightarrow{-I} A \tag{11.13}$$
$$X \xrightarrow{+F|I} ? \tag{11.14}$$

then—whether this is beneficial or harmful to any party depends not only on its being shared $+I$, but on whether it is received and understood correctly $-I$, and on how it is used $F|I$. There is bound to be a variety of social impacts as a result of any shared information.

What may be more important than whether information is shared or not, is how society prepares for the impact of this information getting into the hands of any agent. There is a scaling issue too: macro legal restrictions might be appropriate on the scale of society, but completely fail to protect an individual. So, for every benefit of open information, there might be harm. Across the world, as most societies limit information to some extent. What matters in a society is how those in power regard information, and how justice is served[106].

Trust is an essential glue in cooperation that cannot be eliminated without also eliminating society itself. The potential harm of a well-intentioned technology may well lie in the naivete that spawns it. In this case, what does giving up trust say about someone? It seems clear that, in the wrong hands, blockchain could quickly become harmful. We expect that the immutability of blockchains will prove to be a liability, on a number of levels. The first is in terms of resources consumed. A mechanism for forgetting transactions needs to be built into any system, if it is to scale and be assured of a continued existence, otherwise the cost of data and processing will grow more than linearly over time, until it collapses under its own weight. On a more sinister note, the inability to forget and to forgive the past is the basis for blackmail, extortion, and any number of destabilizing misbehaviors.

Societies may not have the political maturity to deal with such immutable information[107]. Records of thoughts and actions have been preserved into familial and tribal grudges over many generations, leading to much violence. A technology with the ability to forgive and forget is not something anyone has so far seen fit to invent.

CHAPTER 12

THE FUTURE OF MONEY

In his excellent book on debt[Gra11], Graeber points out that modern monetary economics has only existed for a handful of decades, with the full spectrum of instruments we now consider to form Global Economics and the Monetary System. To judge its success or failure on a historical timescale is thus premature. Societal practices take time to equilibrate, and we can expect many more innovations in monetary practice to come, especially with the rise of FinTech or financial information technology. The focus of our book has been to highlight the role of money as a network information technology, which seems to be a topic that has fallen below the radar of economists previously. Our thesis has been that promises have always played a central role in the story of money—an even larger one than debt. Promises relate to the most basic question of trust between individuals and groups.

12.1 TRUST

As we propel ourselves into the information age, the rising torrent of information and communications is leading to a general undermining of trust, particularly in those bodies that act as representatives on behalf of groups and special interests in society. The Trusted Third Parties that have made society work in the past, like governments, institutions, and banks no longer control the flux of information from a position of monopoly or privilege. This opens them to greater scrutiny and empowers individuals with the suggestion that individuals might now do a better job without the coherence of institutional control. It's a perfect revival for *laissez faire* thinking.

In fact, rather than democratizing access to factual information, and levelling the playing field of fairness, the Internet has been used effectively as a loudhailer for mass suggestion, 'fake news', and the marketing of an ever growing pantheon of individual interests, including state propaganda, corporate propaganda, and even ad hoc rumours and gossip that resonate around the echo chambers of segregated social media groups[108]. This has implications for money too.

The downgrading of trust applies to all of the important institutions of our societies, including those that enable a monetary system to be sustained: governments (whose unkept promises are ever easier to discover, and whose singular power can be increasingly wielded by brute computational force), news agencies (whose biases are easier to discover), banks (whose corruptions and misdeeds are easier to discover), and so on. As we have mentioned throughout the book, many aspects of monetary policy are guided by beliefs and guesses about an economic network that teeters in a state of quasi-stability. The 'invisible hand' of Adam Smith, which imagined a benevolent quasi-deterministic emergent outcome from the autonomous actions of agents in a capitalist economy, is a fiction that actually requires calm and thoughtful restraint, not the 'live and let die' mantra of laissez faire. It relies on essential stability of the boundary conditions, and that stability is all too easily wiped out by resonant feedback in the system. Trust needs stability, but too much information undermines that[109].

This gradual undermining of trust in institutions and Trusted Third Parties is potentially catastrophic for cooperative civilization, as we have understood it thus far, and money is at the heart of the matter[110]. Money's inability to maintain fair access to and distribution of what ought to be the shared resources that bind civil society together, is a serious issue—as serious as any other resource management issue, like access to healthcare and housing. Uneven distributions of supply and demand in space and time cannot be instantaneously evened out by economic wishful thinking. A modern theory of economics needs to take this into account.

12.2   MONEY TECHNOLOGY

Our goal here is not to address every defect of economic modelling, but it is appropriate to ask a simpler question. Is there a way in which money can play a role in restoring the confidence to protect a way of life we covet? Is there, for example, an improved technology that would help?

The cryptocurrency response to the 2008 financial crisis was a simple—if naive— attempt to address a perceived problem in the distribution of power using technology. It was one example of how technology can make a difference—but arguably ended up becoming the very monster its creator was trying to counter. The blockchain technologies that came out of that experiment have offered a variety of ways of applying a unique form of public voting on the acceptance of certain transactions, in order to decentralize power. Guidance, by majority decision, is an ongoing political theme in the West—but, if we examine it critically from a systems perspective, it has many flaws and instabilities, amongst which are that it is a blunt instrument poorly suited to quick or targeted responses based on accurate information. For many, the idealism of blockchain held great promise. Instead its dominant feature seems to be becoming a new environmental catastrophe, expending massive energy cost and encouraging criminal activity in the gaming of its method of new money creation[111]. We don't want to disparage the cryptocurrencies unduly, but they serve as a lesson that technology can only be a part of the solution.

In Asia, consumer monetary technology has come much farther than in Europe or America. Consumers can pay by a variety of electronic means, including by simply showing their faces for facial recognition. This goes a long way to freeing economic activity. One of the major hindrances to economic development around the world is the inability to easily pay for goods and services. Many citizens don't even have access to a bank account, locked out by credit ratings in a catch-22. Today, there seems to be no valid excuse for governments not to issue an official bank account to every citizen at birth. Access to banking should be granted as a basic right—this is government's role.

On top of this, communications and transactions are growing in volume and thus also in speed. Stock trading is already faster than human beings can perceive, and soon this will apply to all money. We are at risk of losing control of money. The issue is not a problem with technology: automation can handle the speed, but humans can't. Is it morally and socially right to hand control of money over to machinery and automated systems? So far, all such software systems have failed to offer a totally safe and secure environment, fit for humans.

12.3 MODULARITY

A classical answer to handling separate needs is to *modularize* a system, breaking it up into smaller parts in order to specialize regions as functions. In Promise Theory, we would say partitioning an agent into subagents. In some fields, like Information Technology, modularity is considered a kind of panacea to solve all problems, but there the placement of boundaries can have mixed outcomes. Consider the way land has been diced and sold into private ownership in some parts of the world, preventing sustainable agriculture. To be sustainable and ecological you need a field of cows and a field of grain, to swap each year, but if you only own one or the other you suddenly need to pay for the services of the other, leading to workarounds that may have unwanted side effects.

Money is becoming increasingly modular, thanks to the rising importance of microcurrencies, but society is not fully prepared for this development. Governments are not able to tax these private monies easily, and their accountability is even less transparent than with private banking. One example of modularity would be to separate currencies by scale. The reversal of a trend towards globalization in recent years, marked by increasing nationalism, border protections, and tariffs, etc, may offset this problem temporarily. The principle of scaling of agency, in Promise Theory, suggests that money could be stabilized by separating currencies by scale (see figure 12.1). Autonomous regions with political autonomy can be managed separately to distribute money according to a functional policy, but foreign exchange can disturb this, because the trust relationship between countries is a logically independent issue to the trust issue between local agents and their national institutions.

The benefits of consistency in money clearly (and unsurprisingly) insinuate societies towards a global homogeneity, with a single effective currency, and common politics over the scope of economic interaction (though this does not imply the eradication of multiculturalism). What stands in the way is largely the other side of politics: the fear of losing identity, the desire to maintain ownership, to reject change, etc. Ironically, one of the hindrances in the West to adoption of new technologies that preserve identity is a fear of losing *privacy*. This is another sociological attitude that is likely to evolve over time.

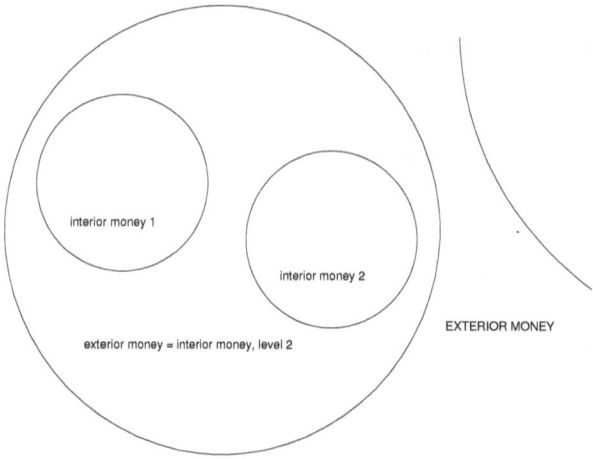

Figure 12.1: Separation of scales suggests different monies for different boundaries.

The future of money cannot be separated from anthropic concerns. It would be naive to think of economics as something separable from politics, given money's roles as a surrogate for trust. Will we even need money in the future? The answer seems to be yes, in spite of the speculations of science fiction authors[112]. However, we clearly do not need the physical proxies of coinage and promissory notes. Will we need banks in the future? That is less certain. We can already see the role of banks being absorbed by other institutions and enterprises—even as banks fight back by trying to offer other services, such as being an authority for *identity*. However, this may well be a problem. We have shown that, without regulation, financial money creators become a privileged class able to act with impunity. However, there is no reason why banks' intermediate role could not be eliminated entirely and the licence to create limited amounts of money be completely distributed amongst societies. Banks would simply be routing hubs, and access to credit would be at the touch of a button. If this sounds like science fiction, then science fiction is already happening through social media platforms, especially in Asia, where electronic payment is some years ahead of the West.

Today, firewalls and private communications technologies enable agents to receive only the information they want to receive. The mixing and moderation of opinions, political and otherwise, which is the key to the semantic stabilization of society, is ironically less effective because of the possibility of discriminating by rich data, leading to targeted semantics of messages. Money's great advantage is its stupidity and lack of allegiance to any particular flag. It is blind to distinctions. In the past countries have employed common or preferred currencies for certain trades (the Dollar, the Euro, etc) because, in the pre-information age, it was expensive to book-keep all the conversion information, or it was politically expedient for the currency owners to control the conversation[Var15]. Now it is both cheap and plausible to have as many currencies as we like. This has political as well as economic consequences. The power of the Internet in managing information suggests that we might be able to monitor the money supply and account for trust in new ways. Will currencies like Euro and Dollar become obsolete? This also seems likely, in the manner of boiling a frog. Global trade needs global currencies, while local distribution needs local currencies. This scaling issue is hard to avoid.

What semantics would we ask of a new kind of currency? What kinds of questions could it help to answer? For now, we shall leave this question for readers to ponder. In this overview, we have focused on the basic semantic properties of money, laying out common interactions between agents in an economic network. We can use the principles of cooperation, described in promise theory, between initially autonomous agents to address these concerns, but we must defer that goal, and the vast subject of economic dynamics, for a sequel.

12.4 NEW PROXIES AND NEW SEMANTICS FOR MONEY

We have distinguished between money and its various representational forms and proxies. We showed that is is possible to describe money consistently and invariantly using agents and promises, without referring to relativistic notions like value. Moreover, we've suggested that keeping these two ideas completely separate would avoid many misunderstandings. Like the metre, the kilogram, and the second, money is thus important for capturing the invariant aspects, and separating them from relativistic and observational distortions. Agents may assess value, but this has no influence on money.

We showed further that, unlike energy in physical science, money is not naturally conserved. Like any promise, it is conserved only to the extent the promise to account correctly can be kept. Money can be created and destroyed, somewhat like energy, but debt and created money are not symmetrical. The semantics of money may persist even after money has been destroyed. Conservation may be compromised by loss of cash, by rounding errors, and by fraud.

We showed that money plays a key role as a network technology, connecting agents with price messages and exchange messages, and that banks act as routers that calibrate money only if regulated by a central authority. Entities from countries to corporations erect firewalls and hubs for money and investment, just as they do for information. Money cannot be understood without a notion of location and time. We've shown how borrowing money can allow agents of overcome obstacles in space and time, and that paying back debt is less important that keeping money well distributed throughout a societal network. The charging of interest on loans is not an effective way of keeping money well distributed, and may even cause new economic obstacles. The success of using negative interest to encourage wealth dispersion is too new to assess yet. The concept of interest is inherently economically unstable, so it requires continuous detailed balance. Eventually, this may be done by automated processes, such as 'artificially smart' systems in order to reach the necessarily level rate of correction. The process could also be decentralized, in principle.

The paradox of money is that the very properties which make it a universal lingua franca for sharing, also allow its penalties to spread virally across those it potentially serves. The fungibility or lack of semantics on money make it fair in opportunity, but blind in its inability to respond in proportion to context. The 'free market' argument for rejecting constraints is balanced by a rise in the number of targeted currencies with specific semantics. In an information age, it will be hard to control the rise of a new ecosystem of currencies, with tailored semantics. Relying on money alone to map out the economy, could lead to a fragmentation of society into special interest groups, while simultaneously allowing privilege to ride roughshod over the remainder. The current design of the financial system makes it trivial for wealth to be siphoned from the bulk poor to the already wealthy. It is known from the mathematics of graphs that the only way to equalize the distribution of transactions in a network is to pump money from richer regions to poorer regions [BBCEM10].

The fundamental asymmetry between money and debt cannot be explained without the accounting of semantics. Even today, simple monetary debt carries more semantics than money itself, as a look into history confirms[Gra11]. The creation of money along side debt is not like the creation of matter and anti-matter, from a virtual reservoir of energy funds. Although banks create money in a superficially similar way, they do so asymmetrically.

The status of cryptocurrencies, based on mobile ledgers, like blockchain is open too; their semantics depend on the specific promises they make concerning what is kept in those ledgers. Clearly, mobile money which can never forget its history is problematic both in terms of the loss of contextual meaning associated with the information (which can only exist relative to the memory of the environment in which it operates).

12.5 THE LANGUAGE OF TRADE IS PRICE

We've shown that buying and selling is an information exchange and 'markets' are information channels that communicate with imperfect information. Selling commodities and specialties are quite different processes, that involve spacetime averaging along different dimensions (analogous to Bayesian and frequency averages). Prices cannot currently represent arbitrary information, and thus the efficient market hypothesis cannot be true. The settling of prices by rational equilibria, as in Game Theory, may be plausible, but only over long timescales much larger than a single trade, and assuming that prices are approximately constant over that timescale. This means that conventional economic arguments can only apply to commodities. Although these limitations exist today, future technologies could actually realize them more plausibly.

We showed that a market price distribution is fundamentally different from a bartering price, and that there are two kinds of price adaptation mechanisms, which we may classify as cognitive and ensemble pricing. These correspond to specialized goods in small markets (where personal 'salesman' negotiation plays a major role in selling) and commodity goods in large markets (where cost of sale is marginal and negotiation is 'take it or leave it'). Value and price are basically unrelated, or at best weakly coupled on a time-varying basis. Marketing is an informational side channel alongside prices, for communicating detailed semantics. We can find little evidence that rational game theoretical equilibria play a significant role in the dynamics of markets.

Money is not the language of trade, in spite of what people sometime chime (money carries little information), rather prices are. Prices may be reliably compared by a common standard measuring stick, which is a key role played by money. Having standard money as a common alphabet, or language of exchange, allows intentions to communicated faithfully and ultimately money to be conserved. If money ceases to be fungible, this acts to constrain what can be purchased for the tokens and thus effectively reduces the connectivity of the trade graph, partitioning and isolating product spaces. In the future, we can expect to see a deepening and a broadening in the nature of price and how it is communicated. A simple numerical tag is no longer the way we review purchases—the semantics of promised exchange now has new channels that are finding their way into our lives through 'smart devices' and services.

The opportunities for smart pricing have hardly been tapped, but (at the time of writing) smartphone 'app' technology is quickly taking over as the new proxy for money, with smart semantics built in. We won't speculate as to the location of that software in the future, but prices are likely to transform from a passive form of signalling to an interactive process soon. Prices on all kinds of goods and services can automatically adjust to discount purchases, or to incorporate and pay taxes transparently—both alleviating consumers of the burden and relieving them of the psychological stress of seeing the details of where their money optimally needs to go to maintain their lifestyle in the long run. There are questions about the perception of fairness and even discrimination law in the concepts of smart adaptive prices. Currently the West is more resistant to innovation where privacy concerns are involved than Asia, but attitudes are likely to change as we get more used to living in a world of information and surveillance.

12.6   UNIVERSALITY, RIGHTS, AND SOCIETY

In any policy based system there is a tension between wanting universality—the same rules, conditions and laws for all—and needing to discriminate and adapt to context. Today, we have very different attitudes about rights, privacy, and boundaries around the world, and yet that single issue lies at the heart of so many conflicts and so many 'inequality traps'. What we perceive to be a just apportioning of basic 'rights' or privileges continues to be a major bone of contention in civil society, and money plays a central role in mediating these matters, given its role as a premier network technology[113].

The law of preferential attachment in interactions implies that rich agents will always be likely to get richer without an executive process to stabilize that inequality. We know from network science that complete equality and complete inequality are both death knells for a system[114]. Similarly, building a fence around an initially 'free' asset, claiming it as property, and doubling down to defend it is a persistent instinct in all animals—and it continues to bedevil our efforts to build a sharing society that works satisfactorily for all. This applies not only to land property. Patent law, for example, continues to be manipulated by wealthy actors at the expense of poorer actors—even while there is a parallel effort to encourage 'Open Source' technologies that are free for anyone to pursue[Mee17]. A richer language for interchange, with even simple permissions, could go a long way to sharing fairly (whatever we decide that to mean) and managing boundaries, assuming everyone would trust in a common system—this we know from the design of rather simple societies in multi-user computer systems.

The non-linearity of money now means that we chase profits from money itself. Money can make money simply by intricate network amplification. Equally, it could all crash down a sinkhole in a relative instant unless regulated. The much lauded freedoms of Western society are greatly overstated when it comes to civil stability. There has to be discipline (i.e. constraint by regulation) to prevent collapse of the dynamical system. Freedom has mainly worked in the past because—given the opportunity of freedom— most people choose to build a fence around a small property holding and never explore the rest of what's out there; similarly, they will keep to a narrow ideology and oppose others, thereby walking a rather narrow and sparse path. This has a natural pruning effect. Semantics are one route to selection (pruning combinatoric growth[115]) in order to keep it approximately linear, detailed balance on the remainder is another.

What we do know about human society is that everything is accelerating. What used to take a generation to change now happens in the blink of an eye. Our laws and systems need to adapt much more quickly to survive this development in the future. Money is the simplest of communicative forms that traditionally abhors semantics, both because semantics add expense and complication, and because limiting access to money is like cutting off one's own hand. Everyone benefits in a network of a certain size—and partitioning to limit access may have grave and unpredictable consequences.

The simple semantics of money are both a help and a hindrance to improving the economic network. Money such as ownerless cash, which cannot alter semantics, without side channels, is easily laundered, but not easily directed fairly. On the other hand, even partially distinguishable money (like BitCoin), which can remember its origins, may not be desirable to some agents, and may be discriminated against on a variety of bases (section 7.16). One can imagine new 'microcurrencies' that are contained within specific regions as a compromise between the desire to keep track of information, and the desire to keep it simple.

In our formulation, the stability of an economic network is a graph theoretical problem, with dynamics of interfering timescales. This makes monetary networks non-deterministic in many ways, leading to a theory that is in line with the modern theories of physics in the 20th century, e.g. statistical mechanics, quantum mechanics, information theory, etc.

When someone proclaims 'it's all about money', it is analogous to saying 'it's all about the Internet'. But money itself is only a causal factor through its sufficient supply. It is a uniquely passive interloper between the semantic forces at the endpoints (agents) of the network. A lack of money in circulation today is like a lack of taxis or trains when we want to travel. We don't need to have our own vehicles, but we crave access when the need arises. It is not without deep anthropological semantics, but it has reached a convenient level of separation of concerns in the modern capital economy. Money links and bind agents through networks of iteratively maintained trust relationships. The largest question of all for the future is: where shall we place our trust?

In order for money to become a liberating force, rather than a tether, its basic design needs to change from macro to micro, in a smart and adaptive way. For all the bluster about interest rates, inflation and GDP in the media, the machinations of our national and global economies have little to do with the needs of ordinary people today. It is surely a goal to address this deficiency in a future design for Money and Ownership—restoring more individual agency to common people. The argument that civil society is about participating in a network is significant. Scaling that network to the entire population of the modern world will be a great innovation. Money, as we recognize it today, could be called a version 1.0 model for society. It is a blunt instrument, predating the information age, with only weak semantics. The information society should be able to do better, if it can confront the issue of even greater scale and speeds. A topic of this size is surely one for future work.

CHAPTER NOTES

NOTES

[1]Our Promise Theory is not to be confused with some earlier attempts at defining promises in Law and Philosophy [Owe06, She11, Sca90, Ati81, Car84]. To readers who are new to the Promise Theory of [BB14a], we advise against taking the term 'promise' in too human or sociological meaning. It is a formal abstraction of intent, which can be made by humans or by proxies, requiring essentially only a labelling of semantic properties.

[2]In the promise theory of [BB14a], many of the traditional tenets of philosophical and legal conceptions of promises were rejected due to their tendency to advance the primacy of obligations. This view is unhelpful both on philosophical and practical grounds, as it injects assumptions of morality where none are needed. The traditional view of promises as generating moral obligations is only a special case that brings more problems than solutions.

[3]Physicists would turn their noses up at this, then we would contend that a theory that intentionally avoids observation for the sake of distaste is of little use.

[4]In [Gra11], Graeber possibly overplays the concept of debt at being central to the understanding of money. Here, we contend that a theory based on promises is simpler, and pointing out that—while these two positions seem initially complementary—debt and money are not dual, because debt carries with it semantics that money does not (see section 6.2). This view seems to be compatible with Graeber's, and clears up some of the contradictions in his account.

[5]A telephone promises to connect us in spite of lacking the free will or cognitive wherewithall to satisfy philosophers of its claim to intentionality.

[6]See the extensive discussion in [Bur13b].

[7]It is worth mentioning that this assumed immediacy, assumed in neoclassical economics may violate several laws of physics in careless application.

[8]We assume there are no minds sufficiently complex to allow complex numbers, however utilities may be tuples with multiple dimensionality.

[9]Some clever accounting is needed to preserve this model of causation, but it works well in its detail. Readers who are familiar with arguments about the causation in inference, should not be confused by physics' pragmatic approach to preconditional ordering.

[10]Scaling is a well defined measure of aggregate complexity in both physics and computer science, often called 'Big O' scaling..

[11]Readers are referred to the discussion of tenancy in [Bur15], which may be considered the inverse of holding.

[12] We could choose a different convention, e.g. owner $= \emptyset$, but self-ownership is a useful short-cut to what follows.

[13] Readers might object on moral grounds to the idea that human agents could be property, but we only observe that human slavery is a phenomenon that exists, however despicable.

[14] Friedman's aversion to collectives[Fri02] seems to take a simplistic (essentially political) view of collectives as being violations of individual freedom, rather than being creative processes, machinery, or combinatoric richness.

[15] This might not be completely true, of course, as human biases in the use of money and choices to employ it may give money the appearance of certain biases, if one tries to eliminate the agencies involved from the description of transactions; however, then one is dealing with a kind of 'effective money' which is encumbered by semantic baggage (somewhat analogous to effective polaron fields in particle physics that are dressed by interactions).

[16] In fact, a theory of money is more like the patchwork theory of local coordinates, analogous the coverage of a metric space by coordinate systems, in Riemannian geometry, with its attendant meaning in a relativistic sense. We shall define it in this way, and make a clear separation between money and the perception of value.

[17] Economists might say the assets have to be in a liquid form, though liquid goat should not be taken to mean a kind of goat flavoured energy drink. Liquidity refers to the ease of flow from owner to owner.

[18] Here physicists would tend to use the term 'quantum numbers' for the analogous role of 'types'. We know of no dairy quantum numbers that would work in this case, however.

[19] In the modern world, we are used to the notion of coins and notes as being traditional money; however, coins were apparently introduced as an innovation only after ledger based accounting for things was used as a form of money, and notes came much later[Gra11].

[20] This specific requirement may be disputed when distinguishing money types is cheap and easy, such as in the information age (see the proposal 2).

[21] Money should have a limited accuracy in terms of denominations. This helps to reduce the amount of compressible and unnecessary information that has to be accounted for. It is this advantageous to make integer money rather than real valued money, because the latter would require infinite information, even though it seems at times as though the number decimal places can be significant.

[22] In modern cloud computing technology, we might call this Mattress as a Service

[23] In antiquity, coins were made of gold or silver and clipped to reduce their value.

[24] The converse type of deposits are sometimes called 'sight deposits' as they are immediately available.

[25] See [BB14a] for the definition of an assisted promise.

[26] There is a nice discussion of this by [Har00], and a different version in [Bur13b].

[27] Indeed, the term value is a 'loaded' one, and is often used strategically to pull the wool over people's eyes about what they are getting in transactions.

[28] Cox has argued that a distributed weakly coupled system of money would be cheaper for society than our current system of money creation, because the strong imposed obligation model of paying interest (whether freely accepted or not) is costly and brittle because it imposes string coupling on the financial system[Coxb, Coxa]. This seems to be one of the goals of BitCoin, however in that model money creation is not free: there is a cost in terms of computational power expended.

[29] The term 'fair' is a difficult one to define. It is a moral judgement, not an objective term, and therefore readers should understand it as a relative term. Objectively, there is no fairness, only network relationships.

[30] The imagining of money is a different imagining than the imagination of value, more analogous to the decision to use occupation as surnames (Smith, Carpenter, Cooper, Burgess, etc).

[31] A slight constraining influence occurs when banks deal with one another, through 'competition'. If a single agent tries to enrich itself at the expense of others, that is easy, but competition between banks is a possible source of constraint. It has been claimed that the causes of the 2008 financial near-collapse originated in the deregulation of implicit money creation, by inflating the price of weak assets through auctioning.

[32] In fact the modern versions of these intermediaries are smarter about keeping proxy accounts where transactions can be kept simply, and clearing of transactions can be performed asynchronously.

[33] Indeed, if later the debt defaults, a third party such as the government may step in to simply erase the debt by issuing new money by fiat, or diverting from some other reservoir.

[34] The muddling of information and meaning is a mistake that is frequently made in technological design.

[35] The case of bankruptcy is interesting. Part of the role of bankruptcy law is to eliminate some of the semantics and responsibilities associated with the inability to pay debt, whether long term or short term. It freezes the clock on processes, and sets a new timescale for resolution. In some cases, the timescale is infinite, and there is never resolution.

[36] The implementation of transactions may follow the classical two-phase or three-phase commit protocols, with their attendant flaws. See also Paul B. transactions ref?.

[37] It might seem strange that coins cannot be owned by someone that holds them. One could register ownership of some coins, but since we cannot distinguish one coin from the next, one can only register some amount of coinage, not the actual coins. One could wrap the coins in a container and own that, but that is semantically a different matter. Then there is no guarantee of what is held inside the container, only a promise of an amount once again. So curiously, one can own an amount of indistinguishable stuff, by containment, but not the actual thing.

[38] Notwithstanding the events of the 2008 financial crises.

[39] MB is grateful to Kevin Cox for helping to understand this point.

[40] Many apartment complexes take out shared debt to save banking fees, but of course they keep track of which units owe money.

[41] In the trade war ensuing between the USA and China at the time of writing (2019), it seems to be the agreement on definitions that is lacking for a successful resolution.

[42] In the deontic moral philosophy, payment might be considered an obligation conferred by a trade, but this is not germane or helpful to our version of things.

[43] Indeed, looking at definitions in books and online, it was nearly impossible to find references to payment without the concept of money being referenced, without going back to the time of the Incas, where payment was made in labour[Fer08, Har11].

[44] The efficient market hypothesis in neoclassical economics suggests that complete information about market circumstances is reflected in pricing. Even if this were not a bizarre suggestion, this cannot be true in our Promise Theory model because the transmission of information through any channel depends on both the promises made by both the sender and receiver. So even if total information could be mapped into an alphabet of price levels, we could not guarantee that the information would be heeded by the recipients. See section 8.5.

[45] We shall not address the symmetry breaking of starting the dilemma. This was discussed in section 7.4 of [BB14a] and further in [Bur13b].

[46] Using an energy analogy, prices are like potential barriers to be overcome be kinetic energy of money, i.e. we might thing of prices as potential money, and money proxies as kinetic money. However, as we show, this analogy does not go very much further.

[47] For example, you've been a good customer, so accept this loyalty discount. Sometimes these discounts are given as coupons which can be traded for future purchases to encourage more buying.

[48] The current norm is to use Fractional Reserve Banking in which lenders have to limit loans to a fraction of the current level of deposits .

[49] See also the reasoning in example 37.

[50] Keynes' theory of money was to encourage the cheap access to money in order to stimulate investment in society[Key30, Key64]. He focused on the money network rather than on the market prices, which later become the approach in neo-classical economics. For an astute history, see [Til07].

[51] There are never enough notes or coins in or out of circulation to allow people to keep their savings in the form of cash. In Switzerland, it is however fashionable to keep vaults with large amounts of cash.

[52] No crystal ball can predict prices of goods in the future, or wages, or even employment levels—so dealing in statistics is of limited value to ordinary people. The economy we read about has little or nothing to do with individuals' lives. This is a clear weakness in the models. Yet, economists use quasi-deterministic predictions to model the economy and set policy. The feedback loops are highly non-linear and thus unstable. If buying power is reduced, businesses may have to reduce wages for employees and they have less to spend, which slows down spending as well as repayment of debt. Interest rates therefore have to serve consumers too, not only lenders, else the burden of debt would grow out of control.

[53] Estimating the rate of inflation to adjust an interest rate for a change in prices is clearly an unparalleled act of prescience, which can only given a reasonable meaning as a statistical (econometric) expectation. Prices may rise and fall, but in which market? The price of apples may rise, and that of oranges may fall. What is the rate of inflation then? Clearly the market for complex manufacturing or agricultural items cannot be predicted in any reasonable sense—how could one possibly take into account shortages of raw materials, natural weather conditions, and other matters that determine the market price? But, in the entirely artificial and constrained world of bond markets, there is sufficient homogeneity to be able to make the argument about limited predictability. For that reason, bond markets play a central role in the determination of interest rates.

[54] This is the benison and the curse of globalization.

[55] This point was made by science fiction writer, scientist and philosopher Isaac Asimov in his initially three laws of robotics. His robot agents had to preserve moral values by 1) never harming humans or allowing them to come to harm, 2) always obeying human wishes, and 3) protecting themselves (since they are valuable commodities). Later, he found the moral inconsistency of these individualistic rules, and added the zeroth law, that humanity (all humans) should not be allowed to come to harm, allowing robots the ability to sacrifice a single human for the good of the race.

[56] The meaning of market is defined quite ambiguously in literature. For instance, we say, 'Is there a market for X?', and 'Y created their own market', implying that a market depends on the existence of buyers; but, then we also speak of market competition and market, implying that a market is centred on sellers.

[57] This identification allows us to discuss noise, distortion, error correction, and capacity, and all the other details attributed to communication.

[58] Star Trek fans: please state the nature of the economic emergency.

[59] In fact economists seem to be quite unable to document clear definitions of anything, however the online website Investopedia does a better job than most textbooks.

[60] Remarkably, in none of the following books [Rap66, Ras01, Dre61, BG02, Sal05, BB06a, Kee11] are the concepts of a market or competition ever defined (if at all to say that the matter is difficult). In [Gin00] there is a minimal formalization of competition, and finally in [OR94] a significant chapter discussing an exchange economy with price distributions determined by games.

[61] Agents do not act as rational maximizers[Kah11, Ari12] on short timescales, and on longer timescales market conditions are likely to have already changed.

[62]In this aspect, economics has something in common with quantum field theory on a graph.

[63]It is unlikely that these distributions would be either flat or multi-modal, i.e. have more than a single peak. If there were several modalities, it would make sense to divide the market into two different markets (see figure 8.1). If the price level were flat, it would be because the product was not sold at all $N_a = 0$, $\forall a$.

[64]With an excellent information technology, one could imagine sampling the prices directly using the previous timelike method, in which case that result could be used.

[65]In a continuum limit, it might be expedient (for the purpose of calculation only) to look at approximate integral representations; however, we shall avoid this here, since the fundamental situation is discrete.

[66]For distributions this is really of the form $\int (dy) L_-(x, y) L_+(y, z) = \delta(x, z)$

[67]This was discussed in section 2.9-2.11 of [Bur15].

[68]As in quantum mechanics, the linear form allows us to project the influence of our incomplete information into a set of local eigenstates at different scales. Since one can never really obtain a complete 'God's eye view' of the network, with perfect information, we have to be content to live with this probabilistic description, and gauge its stability.

[69]The impartiality of many regulatory agencies was questioned in the wake of the 2008 financial crisis[For16], and several so-called trusted institutions have since been indicted for institutional fraud.

[70]We speculate that this is often because taxi drivers are often poorly paid, part time workers trying to avoid oversight and subsequent taxation.

[71]See the review in [Mir89].

[72]It's like the truism that the best time to predict the weather is when it stays the same.

[73]Liberalism and discrimination-free reasoning are luxuries of sparsely populated systems with plentiful resources, so that there is limited contention. As systems fill up to a critical capacity, contention increases and the pressure to discriminate between agents along arbitrary distinctions grows. The morality of a plentiful world is usually quite different from the morality of a scarce world.

[74]What turned out to be missing from both physics and economics were essential arguments about *scale* and *indeterminism*[Mir89], i.e. the role of large number approximations and loss of distinct information about a system (entropy). Keynes was apparently the first to point to the need to take uncertainty and indeterminism seriously[Key30, Key64], but his work was only partially heeded and later misrepresented for decades for political purpose[Min82].

[75]Simmel discussed the extent to which the acquisition of money became a goal in itself and extended one's social identity[Sim11].

[76]The behaviours of the macro-economy on large and long scales is interesting, but not very practical, since human lives focus on payments and transactions day by day, not decade by decade. It is far from clear that understanding the patterns of macroeconomic behaviour would help in the determination of any policy decision on the scale of human lives. Even insight to avoid future catastrophes or 'shocks' is not guaranteed by such a study, if we seek to guide human behaviour. The sudden catastrophes of the financial crises in the 70s, 80s, 90s, 2000s, etc, were all more sudden than the implicit timescales of macroeconomics, and their causes were in no way represented or diagnosable by the models. If we are to approach an improved understanding of the transactional and abrupt changes in economic systems, in the same way that physics took on such phenomena, then economics must also embrace techniques that do not exclude them by design.

[77]The obfuscation of ideas by terminology in economics seems to be a big problem to the layperson: when there is a lack of principles to steer by, there tends to be an explosion of names for phenomena whose interrelationship is suppressed.

[78]Economists refer to marginal quantities, when meaning derivatives with respect to its parametric variable. Scale-free marginal increases have the form

$$\frac{dX}{X}.$$

(12.1)

This kind of scaling embodies the assumption of scale-free behaviour, which excludes semantic effects.

[79]Even in a world with collective bargaining, wage increases have to be ratified by financing.

[80]German newspapers write a lot about home ownership, and always insist that one perceives a house as an entity which loses nearly all of its value. That is not the case with the land for which they merely assume that the v value is constant.

[81]The normal functioning of a capitalist economy is that debts and their (positive) interest repaid by the rents earned from capital assets. The source of the tenant's money is ultimately borrowing from banks, perhaps supplied via the tenant's wages. Capital is similar to a bond, in that it promises future income.

[82]Although Minsky never formalized his ideas as a dynamical model, in the traditional manner of differential equations, Keen did so[Kee95] and others have expanded on these models, e.g.[GH15, GH16].

[83]The methodology thus consists of formulating a small set of coupled oscillator equations, and searching for the stable normal modes of those coupled oscillators. For stability analyses, one can search for the attractors, and Lyapunov exponents, in a way that is not practical in a microscopic model. Interpreting the equations in the context of the real economy is a matter of some speculation, however, given the low level of detail in such a mean field approach.

[84]A semblance of stability can be maintained by timely corrections as a matter of public policy, on a timescale at which the system remains almost linear. This is essentially like making frequent course corrections to a spacecraft trajectory as it passes close to unexpected gravitational bodies. If the corrections are timely, a stable trajectory is maintained. If one waits too long, the craft may be lost forever. The question we have not yet answered for economics seems to be: what precisely are those key timescales in an economy?

[85]We shall use the notation of Grasselli and Huu (GH). There is some poor typography in [Kee95], and GH consider a more general case with greater lucidity.

[86]The attempt to build a causal model that makes the links from micro to macroeconomics has acquired a bad reputation, in some circles, because of the attention afforded to 'demand curves' and the attempt to extrapolate from a Robinson Crusoe economy to a society (see a discussion in [Kee11]). However, that extrapolation was only based on wishful thinking, not on a mathematical model or formal reasoning, and it seems frankly one of the more bizarre fabrications of economic reasoning.

[87]A further delay in the action of the equations comes from real-life *hysteresis*. Because of the normal periodic accounting, the response to failures is likely to be delayed by at least an accounting period. There is thus natural inertial lag in the system, which depends on the particular practices of an agent that depends on another. The main argument for discounting hysteresis is that bulk time-based interference smooths out such effects. This means that one has to be dealing with a diversified portfolio of equivalent interactions, in which there is no fixed time correlation. This suggests only economies of above a certain size will be modelled well by a Goodwin model.

[88]The name derives from the well-known cryptographic method of cypher blockchaining (CBC)[NIS80], in which the encrypted result of a previous block is used as the cipher key for the data of the subsequent block, forming a chain whose informational entropy[SW49, CT91] increases monotonically with the number of blocks. The data structure is said to be a form of Merkle tree.

[89]Many claims have been made about the blockchain and the various applications that have come to be built upon it. The clearest exposition of its motivation and intent is probably still the original paper by

Nakomoto[Nak08], but there are also plenty of reviews and discussions clarifying its implications (e.g see [Kan17a, Kan17b, Kan17c, Bul16]). Our aim here is to offer a promise theoretic perspective, which underlies the causal factors and their implications, to look for the principles amongst the many promises that are claimed for blockchain. We therefore examine the elementary assumptions, and evaluate some of the claims.

[90]We recommend that readers use these notes in tandem with a clear overview of blockchains offered in [BSAB$^+$17].

[91]Some notation used in this chapter:

1. N_ℓ is a node running software agents for blockchain system, at a location ℓ, where ℓ is a subscript representing a node instance location, which might host several human accounts.

$$N_\ell \xrightarrow{+\text{SW}} *. \tag{12.2}$$

2. $S = N_S$ is the node from which the transaction is originated, i.e. the user-account node.

3. $R = N_R \in N_M$ is the node which ends up mining the accepted block containing the transaction. A node cannot promise the role of a sender or a receiver without having at least one PPkey pair:

$$S_i = A_i \xrightarrow{+S\,|\,\text{PPkey}_i} * \tag{12.3}$$

$$R_j = A_j \xrightarrow{+R\,|\,\text{PPkey}_j} * \tag{12.4}$$

By throwing away all of its private keys an agent ceases to be a participant.

4. $N_M \equiv M_\ell = N_\ell$ is an alias for the subset of nodes which promises mining the blocks β_α to include the transaction

5. N_W is the node on which a user's wallet exists.

6. B_σ is block with sequence number $\sigma = 1, 2, \ldots$, accepted into the blockchain, into which a number of transactions is incorporated:

$$B_\sigma \xrightarrow{\{\tau, \tau', \ldots\}} * \tag{12.5}$$

Block promises are in scope of all agents.

7. $C_\ell = \{\beta_1, \beta_2(\beta_1), \ldots \beta_\sigma(\beta_{\sigma-1}(\beta_{\sigma-2}(..)))\}$ is the blockchain replica at location N_ℓ.

8. $\mathbf{N} = (\{N_\ell\}, \{\pi_\ell\})$ is the superagent comprising the total peer consensus network. This is a superagent of nodes and the software promises that bind them to the ledger system[Bur15].

[92]Indeed this is encouraged as part of the transactional security of BitCoin[Nak08].

[93]The common rhetoric that blockchain is a trust-free technology is overstated and quite misleading. The aim of the design, however, is to be able to detect when nodes do not keep the promises of the software (by use of cryptographic hashes), so that the trustworthiness is somewhat verifiable. In practice this requires all agents to trust a single calibrated source for the software, thus the software origin becomes the single trusted party. Trust in its integrity is built on the transparency of the software and its promises.

[94]The distributed consensus here refers to the promises about stored data. It does not necessarily apply to other promises, such as software promises, versions, or other resources. Note also that a blockchain is not a 'journal' which can be rolled back and replayed as in other databases[BC11].

[95]Getting the same data in the end is relatively easy. Getting the order of a sequence with distributed (non-calibrated) origin is much harder, so ordering invariance is the reason for the linearization criteria

above. Conflict free replicated data types - merge all time lines (like git/version control) requires a notion of orthogonality of changes. Consensus means to engineer only one correct outcome.

[96] A sharding of a service is a partitioning into non-overlapping sub-services, each of which covers its subset independently, for parallel scalability.

[97] Causality assures that data, where order is important, arrive at a single location in a unique FCFS order. This would not be possible for more than a single node, without some form of memory of previous data. So, in this approach, arrivals are treated as a first order Markov process, i.e. a memoryless transaction process. This approach is relatively fast, because the semantic selection is determined by the election of a master node, and the decision is cached and reused for multiple transactions. If data arrive quickly, this is an advantage, because serialization through a fixed point makes it $O(N)$ in the length N of the queue.

[98] This is a *masterless* system. It has no single point of validation, and achieves consensus by *equilibration*[Bur12a, Bur13b] based on linkage to previous states of consensus (hence a chain). There is no causal ordering, only semantic ordering, so we cannot be sure when equilibrium will be reached. However, since blocks are added sequentially, equilibrium of past blocks gradually decouples from the ordering of newer blocks, allowing a gradual increase of trust in stability of the data. This form of data chain is the exact opposite of a Markov chain: each step depends cumulatively on a growing memory of past transactions.

[99] There is a similar situation in thermodynamics, To unpick the entropy of mixing or equilibrium in a large reservoir of transactions with different labels, and order them into a stream, 'Maxwell's Daemon' examines every single packet by brute force (a master server) and separates them by their state or temperature. The work cost is very high, rescuing the second law of dynamics from potential brute force attack.

[100] A ledger is a partial ordering of items. A chain is the simplest such structure, with maximal fragility. Tree structures, or directed acyclic graphs (DAG) generalize the properties of chains, and have already come into use within 'blockchain' systems, through sharding, etc[BSAB+17]. The degree of decentralization in the operations performed on ledgers is therefore a continuum of possibilities that range from complete master-node centralization with pre-ordering, to complete peer decentralization with post hoc ordering.

[101] Analysis of chains in [BB14a] concludes is that the number of promises, explicitly written, to ensure precise promise keeping through a dependency chain grows as fast as $O(N^2)$ in the number of agents (e.g. the number of blocks in a blockchain). The cost of adding a block might be high $(O(N))$ or low $(O(1))$, depending on how the promises are kept, but the potential vulnerability to failing to keep the promises of the full chain grow like N^2[BB14a].

[102] We shall not try to evaluate this belief here (see [Bur4]), only point out the consequences of the belief.

[103] Assembling a body of independently verified evidence is the essence of the scientific method, and the Gaussian theory of statistical uncertainty.

[104] Ostensibly out of the hands of untrusted human clients, though this assumption turns out to be naive.

[105] As a networking technology, it resembles the distributed cooperation of autonomous BGP zones, rather than the centralized management of OpenFlow.

[106] One cannot avoid the question of whether the decentralization of responsibility, an apparent dream of both the political left and right, is an entirely misconceived notion, given its inevitable side effects. In essence, using a technology to try to eliminate trust is no less than handing over trust from one elite to another (e.g. from regulated bankers to unregulated technologists).

[107] Governments, where power struggles naturally weaponize information, already relish the prospect of using digital technology to track and profile individuals suspected of threatening opinions and behaviours. Paranoia is a spiralling risk. Already today we confuse culture and nationality with intent: a stamp from the wrong country on your passport, or an incidental image on social media, or making an inappropriate

comment in jest, can make you an enemy of the state. Being in the wrong place at the wrong time can be interpreted as a marker of hostility, because intent is as much inferred by promisees as it is promised by promisers.

[108] One of us, MB, wrote about this possibility in a novel in a 2003, which has turned out to be fairly accurate in its predictions[Bur05b].

[109] See, for instance, [Bur13b] for a more detailed understanding of this point.

[110] The depersonalization of authority has been key to modern civil society, where kin and tribal relations are replaced by third parties that we consider to be more fair. The indications are that we are returning to a more tribal view given the tools for scaling such relationships with electronic communications.

[111] The IT industry creates more emissions than the airline industry today, and the proliferation of unused data in storage probably already dwarfs the plastic waste crisis—yet we view information technology as a clean technology.

[112] In the movie Star Trek, First Contact, Captain Picard famously says 'the economy of the future works differently', implying that money is not needed in the future, yet every episode of the series that deals with such issues involves money in some form or another.

[113] In Promise Theory, the interpretation of 'rights' is very simple and logical. Rights stem from permissions to access resources of other agents. This is an economic matter rather than a moral one, thus the morality of a society is subordinate to its economy. This follows from the causal independence of agents—and while it might be disappointing to many readers, it is palpably seen to be true in all cultures.

[114] Complete equality is a 'heat death' of maximal entropy, while complete inequality is a sink for all resources that ends in starvation of resources. See [BBCEM10] for a technical proof, and [Bar02] for a nicer read.

[115] See the discussion of pruning and fecundity in [Bur19b].

BIBLIOGRAPHY

[ABSB⁺17]   M. Al-Bassam, A. Sonnino, S. Bano, D. Hrycyszyn, and G. Danezis. Chainspace: A sharded smart contracts platform. *arXiv:1708.03778v1 [cs.CR]*, 2017.

[All11]   T. Alloway. The fed can't go bankrupt anymore. Financial Times, 2011.

[Ari12]   D. Ariely. *The Honest Truth About Dishonesty*. Harper Collins, New York, 2012.

[Art09]   W.B. Arthur. *The Nature of Technology*. Penguin, 2009.

[Ati81]   P.S. Atiyah. *Promises, Morals and Law*. Clarendon Press, Oxford, 1981.

[Axe84]   R. Axelrod. *The Evolution of Co-operation*. Penguin Books, 1990 (1984).

[Axe97]   R. Axelrod. *The Complexity of Cooperation: Agent-based Models of Competition and Collaboration*. Princeton Studies in Complexity, Princeton, 1997.

[Bar02]   A.L. Barabási. *Linked*. (Perseus, Cambridge, Massachusetts), 2002.

[bas14]   Basel committee on banking supervision: External audits of banks. www.bis.org, ISBN 978-92-9131-276-4, Bank for international settlements, 2014.

[BB06a]   W.J. Baumol and A.S. Blinder. *Essentials of Economics, Principles and Policy (10th edition)*. Thomson, 2006.

[BB06b]   J.A. Bergstra and M. Burgess. Local and global trust based on the concept of promises. Technical report, arXiv.org/abs/0912.4637 [cs.MA], 2006.

[BB13]   J.A. Bergstra and M. Burgess. A static theory of promises. Technical report, arXiv:0810.3294 v5 [cs.MA], 2013.

[BB14a]   J.A. Bergstra and M. Burgess. *Promise Theory: Principles and Applications*. χ*tAxis* Press, 2014.

[BB14b]   J.A. Bergstra and M. Burgess. Promises, impositions, and other directionals. Technical report, arXiv:1401.3381 [cs.MA], 2014.

[BBC18]   BBC. World business report, April 2018.

[BBCEM10]   J. Bjelland, M. Burgess, G. Canright, and K. Eng-Monsen. Eigenvectors of directed graphs and importance scores: dominance, t-rank, and sink remedies. *Data Mining and Knowledge Discovery*, 20(1):98–151, 2010.

[BBV18]   BBVA. Bbva and indra deliver the worlds first blockchain-supported corporate loan, April 2018.

[BC11]   M. Burgess and A. Couch. On system rollback and totalized fields: An algebraic approach to system change. *J. Log. Algebr. Program.*, 80(8):427–443, 2011.

[BCE04]   M. Burgess, G. Canright, and K. Engø. A graph theoretical model of computer security: from file access to social engineering. *International Journal of Information Security*, 3:70–85, 2004.

[BdL13]   J.A. Bergstra and K. de Leeuw. Bitcoin and beyond: Exclusively informational money. *arXiv:1304.4758v2 [cs.CY]*, 2013.

[Bei05]   E.D. Beinhocker. *The Origin of Wealth*. Random House, 2005.

[Ber15]   J.A. Bergstra. Informational money, islamic finance, and the dismissal of negative interest rates. *Univesity of Amsterdam preprint*, 2015.

[Bet13]   L.M.A. Bettencourt. The origins of scaling in cities (with supplements). *Science*, 340:1438–1441, 2013.

[BF05]   M. Burgess and S. Fagernes. Voluntary economic cooperation in policy based management. *Unpublished*, (2005).

[BF06]   M. Burgess and S. Fagernes. Autonomic pervasive computing: A smart mall scenario using promise theory. *Proceedings of the 1st IEEE International Workshop on Modelling Autonomic Communications Environments (MACE); Multicon verlag 2006. ISBN 3-930736-05-5*, pages 133–160, 2006.

[BG02]   E. Brousseau and J-M. Glachant, editors. *The Economics of Contracts Theory and Applications*. Cambridge University Press, 2002.

[Big18]   John Biggs. Sierra leone just ran the first blockchain-based election. http://tcrn.ch/2GtiaVI, March 2018.

[Bis02]   M. Bishop. *Computer Security: Art and Science*. Addison Wesley, New York, 2002.

[BLH⁺07]   L.M.A. Bettencourt, J. Lobo, D. Helbing, C. Hühnert, and G.B. West. Growth, innovation, scaling and the pace of life in cities. *Proceedings of the National Academy of Sciences*, 104(107):7301–7306, 2007.

[BM11]   J.A. Bergstra and C.A. Middelburg. Interest prohibition and financial product innovation. https://arxiv.org/abs/1104.2471, 2011.

[Bri06]   B. Briscoe. Metcalfe's law is wrong. *IEEE Spectrum*, 2006.

[BSAB⁺17]   S. Bano, A. Sonnino, M. Al-Bassam, S. Azouvi, P. McCorry, S. Meiklejohn, and G. Danezis. Sok: Consensus in the age of blockchains. *arXiv:1711.03936v2 [cs.CR]*, 2017.

[Bul16]   Aleksandr Bulkin. Explaining blockchain: how proof of work enables trustless consensus. Keeping Stock, 2016.

[Bur4]   M. Burgess. *A Treatise on Systems: Volume 2: Intentional systems with faults, errors, and flaws*. in progress, 2004-.

[Bur05a]   M. Burgess. An approach to understanding policy based on autonomy and voluntary cooperation. In *IFIP/IEEE 16th international workshop on distributed systems operations and management (DSOM), in LNCS 3775*, pages 97–108, 2005.

[Bur05b]   M. Burgess. *Slogans: The End of Sympathy*. χt-axis Press, 2005.

[Bur09]   M. Burgess. Knowledge management and promises. *Lecture Notes on Computer Science*, 5637:95–107, 2009.

[Bur12a]   M. Burgess. Deconstructing the 'cap theorem' for cm and devops. http://markburgess.org/blog.html, 2012.

[Bur12b]   M. Burgess. *New Research on Knowledge Management Models and Methods.*, chapter What's wrong with knowledge management? The emergence of ontology. InTech, 2012.

[Bur13a]   M. Burgess. *In Search of Certainty - The Science of Our Information Infrastructure*. χtAxis Press, November 2013.

[Bur13b]   M. Burgess. *In Search of Certainty: the science of our information infrastructure*. Xtaxis Press, 2013.

[Bur14]   M. Burgess. Spacetimes with semantics (i).
              http://arxiv.org/abs/1411.5563, 2014.

[Bur15]   M. Burgess. Spacetimes with semantics (ii).
              http://arxiv.org/abs/1505.01716, 2015.

[Bur16a]   M. Burgess. On the scaling of functional spaces, from smart cities to cloud computing. *arXiv:1602.06091 [cs.CY]*, 2016.

[Bur16b]   M. Burgess. Spacetimes with semantics (iii).
              http://arxiv.org/abs/1608.02193, 2016.

[Bur17a]   M. Burgess. Banks, brains, and factories. markburgess.org 'blog' essay, January 2017.

[Bur17b]   M. Burgess. A spacetime approach to generalized cognitive reasoning in multi-scale learning.
              https://arxiv.org/abs/1702.04638, 2017.

[Bur19a]   M. Burgess. From observability to significance in distributed systems. *arXiv:1907.05636 [cs.MA]*, 2019.

[Bur19b]   M. Burgess. *Smart Spacetime*. χtAxis Press, 2019.

[BYLMKG17]   Y. Bar-Yam, J. Langlois-Meurinne, M. Kawakatsu, and R. Garcia. Preliminary steps toward a universal economic dynamics for monetary and fiscal policy. *arXiv:1710.06285*, 2017.

[Car84]   J.P.W. Cartwright. An evidentiary theory of promises. *Mind (New Series)*, 93(370):230–248, 1984.

[CEM07]   G. Canright and K. Engø-Monsen. *Handbook of Network and System Administration*, chapter Some Aspects of Network Analysis and Graph Theory. Elsevier, 2007.

[Cho59]   N. Chomsky. On certain formal properties of grammars. *Information and Control*, 2(2):137–167, 1959.

[CIW17]   CIW. China third-party online payment market overview q1 2017. *China Internet Watch*, 2017.

[Cla]   Thomas Claburn. Bitcoin's blockchain: Potentially a hazardous waste dump of child abuse, malware, etc. The Register, March.

[Coi]   CoinTelegraph. Goodbye kickstarter? the blockchain-based project aims to challenge the crowdfunding sector.

[Coxa]   K. Cox. An overview of promise theory and financial systems.

[Coxb]   K. Cox. Promise theory and financial systems.

[CT91]   T.M. Cover and J.A. Thomas. *Elements of Information Theory*. (J.Wiley & Sons., New York), 1991.

[Dia97]   J. Diamond. *Guns, Germs, and Steel*. Vintage, 1997.

[Dre61]   M. Dresher. *The mathematics of games of strategy*. Dover, New York, 1961.

[Dun96]   R. Dunbar. *Grooming, Gossip and the Evolution of Language*. Faber and Faber, London, 1996.

[Exc]   Stack Exchange. Upgradeable smart contract question.

[Fer08]   N. Ferguson. *The Ascent of Money*. Penguin, 2008.

[For16]   R. Foroohar. *Makers and Takers*. Crown Business, 2016.

[Fri02]   M. Friedman. *Capitalism and Freedom (40th Anniversary edition)*. Univesity of Chicago Press, 1962, 2002.

[Gal12]   S. Galam. *Sociophysics*. Springer, 2012.

[GH15]   M.R Grasselli and A.N. Huu. Inflation and speculation in a dynamic macroeconomic model. *J. Risk and Finanical Management*, 8:285–310, 2015.

[GH16]   M.R Grasselli and A.N. Huu. Inventory and growth cycles with debt-financed investment. *arXiv 1610.00955v*, 2016.

[Gil81]   R. Gilmore. *Catastrophe Theory for scientists and engineers*. Dover, (New York), 1981.

[Gin00]   H. Gintis. *Game Theory Evolving*. Princeton University Press, 2000.

[GL06]   Jim Gray and Leslie Lamport. Consensus on transaction commit. *ACM Trans. Database Syst.*, 31(1):133–160, March 2006.

[Glo17]   Yicai Global. Tianjin teams up with ant financial to build first cash-free city in northern china. www.yicaiglobal.com/news/, June 2017.

[Gra11]   D. Graeber. *Debt: the first 5000 years.* Melville House, 2011.

[Gre88]   J.A. Green. *Sets and Groups.* Routledge & Kegan Paul, 1988.

[Har00]   K. Hart. *The Memory Bank: money in an unequal world.* Profile Books, 2000.

[Har05]   K. Hart. *The Hit-Man's Dilemma.* Prickly Paradigm Press, 2005.

[Har07]   K. Hart. Money is always personal and impersonal. *Anthropology Today*, 23(5):12–16, 2007.

[Har11]   Y.N. Harari. *Sapiens.* Vintage, 2011.

[Hay44]   F.A. Hayek. *The Road to Serfdom.* Routledge, 1944.

[Hic73]   J. Hicks. *Capital and Time.* Clarendon Press, Oxford, 1973.

[Hub18]   BlockChain Hub. Blockchains & distributed ledger technologies, 2018.

[HW90]   M.P. Herlihy and J.M. Wing. Linearizability: A correctness condition for concurrent objects. *ACM Transactions on Programming Languages and Systems*, 12(3):463–492, 1990.

[Ind11]   D. Indiviglio. The destruction of money: Who does it, why, when, and how? *The Atlantic*, Apr 8 2011.

[Inv17a]   Investopedia. Liquidity. `http://www.investopedia.com` retrieved, June 2017.

[Inv17b]   Investopedia. Money supply. `http://www.investopedia.com` retrieved, June 2017.

[JRS11]   F.P. Junqueira, B.C. Reed, and M. Serafini. Zab: High-performance broadcast for primary backup systems. *IEEE/IFIP 41st International Conference on Dependable Systems and Networks*, pages 245–256, 2011.

[Kah11]   D. Kahneman. *Thinking, Fast and Slow.* Penguin, London, 2011.

[Kan17a]   Erik Kangas. Understanding blockchains (and bitcoin) part 1: Concepts. The LuxSci FYI Blog, 2017.

[Kan17b]   Erik Kangas. Understanding blockchains (and bitcoin) part 2: Technology. The LuxSci FYI Blog, 2017.

[Kan17c]   Erik Kangas. Understanding blockchains part 3: Ethereum, or moving beyond bitcoin. The LuxSci FYI Blog, 2017.

[Kee95]   S. Keen. Finance and economic breakdown: modelling minsky's financial instability hypothesis. *Journal of Post Keynesian Economics*, 17(4):607–635, 1995.

[Kee11]   S. Keen. *Debunking Economics (2nd ed)*. Zed, 2011.

[Kee17]   S. Keen. *Can we avoid another financial crisis?* Polity, 2017.

[Key30]   J.M. Keynes. *A Treatise on Money*. Harcourt Brace, 1930.

[Key64]   J.M. Keynes. *The General Theory of Employment, Interest, and Money*. Harcourt Brace, 1964.

[Kle76]   Leonard Kleinrock. *Queueing Systems: Computer Applications*, volume 2. John Wiley & Sons, Inc., 1976.

[Kru08]   P. Krugman. *The return to depression economics*. Penguin, 2008.

[Kul68]   S Kullback. *Information theory and statistics*. Dover, 1968.

[KW13]   P. Krugman and R. Wells. *Macroeconomics*. Worth Publishers, 2013.

[Lam78]   Leslie Lamport. Time, clocks, and the ordering of events in a distributed system. *Commun. ACM*, 21(7):558–565, July 1978.

[Lam01]   Leslie Lamport. Paxos Made Simple. *SIGACT News*, 32(4):51–58, December 2001.

[Lay06]   Richard Layard. *Happiness: Lessons from a New Science*. Penguin, 2006.

[Lee]   G. Lee. Life beyond cryptocurrencies—hong kong and china fintech firms show there is more to blockchain.

[LP97]   H. Lewis and C. Papadimitriou. *Elements of the Theory of Computation, Second edition*. Prentice Hall, New York, 1997.

[Mee17]   H. Meeker. *Open Source for Business*. Fleming, 2015-2017.

[Min]   H. Minksy. *Stabilization of an Unstable Economy*. McGraw Hill.

[Min82]   H. Minsky. *Inflation, Recession, and Economic Policy*. Wheatsheaf, Sussex, 1982.

[Min86]   H. Minsky. *Stabilizing an Unstable Economy*. Yale University Press, 1986.

[Mir89]   P. Mirowski. *More Heat than Light*. Cambridge, 1989.

[MRT14]   M. McLeay, A. Radia, and R. Thomas. Money creation in the modern economy. Technical report, Bank of England Quarterly Bulletin, 2014.

[Nak08]   S. Nakamoto.   Bitcoin: a peer to peer electronic cash system. http://nakamotoinstitute.org/bitcoin/, October 31 2008.

[Nas96]   J.F. Nash. *Essays on Game Theory*. Edward Elgar, Cheltenham, 1996.

[New03]   M.E.J. Newman. The structure and function of complex networks. *SIAM Review*, 45:167–256, 2003.

[NIS80]   NIST. Des modes of operation. Technical report, NIST, 1980.

[NM44]   J.V. Neumann and O. Morgenstern. *Theory of games and economic behaviour*. Princeton University Press, Princeton, 1944.

[OL88]   B.M. Oki and B.H. Liskov. Viewstamped replication: A new primary copy method to support highly-available distributed systems. In *Proceedings of the Seventh Annual ACM Symposium on Principles of Distributed Computing*, PODC '88, pages 8–17, New York, NY, USA, 1988. ACM.

[Ols71]   M. Olsen. *The Logic of Collective Action*. Harvard Univesity Press, 1965,1971.

[OO14]   Diego Ongaro and John Ousterhout. In search of an understandable consensus algorithm. In *Proceedings of the 2014 USENIX Conference on USENIX Annual Technical Conference*, USENIX ATC'14, pages 305–320, Berkeley, CA, USA, 2014. USENIX Association.

[OR94]   M.J. Osborne and A. Rubenstein. *A Course in Game Theory*. MIT Press, 1994.

[OT06] A. Odlyzko and B. Tilly. A refutation of metcalfe's law and a better estimate for the value of networks and network interconnections. *IEEE Spectrum*, 2006.

[Owe06] D. Owens. A simple theory of promises. *Philosophical Review*, 115:51–77, 2006.

[PBMW98] L. Page, S. Brin, R. Motwani, and T. Winograd. The pagerank citation ranking: Bringing order to the web. *Technical report, Stanford University, Stanford, CA*, 1998.

[Pea88] J. Pearl. *Probabilistic Reasoning in Intelligent Systems: Networks of Plausible Inference*. Morgen Kaufmann, San Francisco, 1988.

[Pik14] T. Piketty. *Capital in the twenty-first century*. Belknap, Harvard University Press, 2014.

[Rap66] A. Rapoport. *Two-Person Game Theory*. Dover, New York, 1966.

[Ras01] E. Rasmusen. *Games and Information (Third edition)*. Blackwell publishing, Oxford, 2001.

[Rif15] J. Rifkin. *The Zero Marginal Cost Economy*. Macmillan, 2015.

[Rus18] B. Russel. Proposed roads to freedom, (ii) problems of the future. 1918.

[Sal05] B. Salanié. *The Economics of Contracts (second edition)*. MIT Press, 1994,2005.

[Sca90] T. Scanlon. Promises and practices. *Philosophy and Public Affairs*, 19(3):199–226, 1990.

[Sch54] J.A. Schumpeter. *History of Economic Analysis*. Routledge, 1954.

[Sea12] D. Searls. *The Intention Economy*. Harvard Business Review Press, 2012.

[She11] H. Sheinman. *Promises and Agreements*, chapter Introduction: promises and agreements, pages 3–57. Oxford University Press, 2011.

[Sim11] G. Simmel. *The Philosophy of Money*. Routledge and Kegan Paul, 2011.

[Smi04] M.E. Smith. The archaeology of ancient state economies. *Annual Review of Anthropology*, 33:73–102, 2004.

[Suz17]   J. Suzman. When a 200,000-year-old culture encountered the modern economy. *The Atlantic*, July 24 2017.

[SW49]   C.E. Shannon and W. Weaver. *The mathematical theory of communication*. University of Illinois Press, Urbana, 1949.

[Tai88]   J. Tainter. *The Collapse of Complex Societies*. Cambridge, 1988.

[Til07]   G. Tily. *Keynes's General Theory, The Rate of Interest, and 'Keynesian' Economics*. Palgrave Macmillan, 2007.

[Tof70]   A. Toffler. *Future Shock*. Random House, 1970.

[Tof80]   A. Toffler. *The Third Wave*. Bantam, 1980.

[Tur69a]   A.R.J. Turgot. Paper on lending at interest. *The Collected Writings, Speeches and and Letters*, 1769.

[Tur69b]   A.R.J. Turgot. Value and money. *The Collected Writings, Speeches and and Letters*, 1769.

[Var15]   Y. Varoufakis. *The Global Minotour*. Zed, 2011-2015.

[Wes01]   D.B. West. *Introduction to Graph Theory (2nd Edition)*. (Prenctice Hall, Upper Saddle River), 2001.

[Wik]   Wikipedia. Net neutrality.

[Woo19]   E. Woollacott. Should ai own their own ip? Raconteur, March 2019.

[Yue17]   P. Yue. Ant financial partners with chinese city to build cashless society. www.chinamoneynetwork.com, June 2017.

[ZLX15]   X.Z. Zhang, J.J. Liu, and Z.W. Xu. Tencent and facebook data validate metcalfes law. *Journal of Computer Science and Technology*, 30(2):246–251, 2015.

[ZSHD04]   W.X. Zhou, S. Sornette, R.A. Hill, and R.I.M. Dunbar. Discrete hierarchical organization of social group sizes. *Proc. Royal Soc.*, 272:439–444, 2004.

INDEX

ACKNOWLEDGEMENTS

MB is grateful to Kevin Cox and Zhaoling Xu for discussions on certain details, and also to Lisa Caywood, Keith Hart, Robert Johnson, Bill Janeway, and Michael E. Smith for helpful comments and references. None of these is responsible for any part of the content.

ABOUT THE AUTHORS

Jan Bergstra is a retired Dutch computer scientist, living in Utrecht. He has worked at the Institute of Applied Mathematics and Computer Science of the University of Leiden, and the Centrum Wiskunde & Informatica (CWI) in Amsterdam. In 1985 he became Professor of Programming and Software Engineering at the Informatics Institute of the University of Amsterdam and Professor of Applied Logic at Utrecht University. His work has focused on logic and the theoretical foundations of software engineering, especially on formal methods for system design. He retired as a full professor in the end of 2016. He is best known for work on algebraic methods for the specification of data and computational processes in general. Jan's current affiliation is Minstroom Research BV, and he can be reached via janaldertb@gmail.com.

Mark Burgess is a British theoretical physicist, turned computer scientist, living in Oslo, Norway. After holding a number of research and teaching positions, he was appointed Professor of Network and System Administration at Oslo University College in 2005, which he held until resigning in 2011 to found the CFEngine company. He is the originator of the globally used CFEngine software as well as founder of CFEngine AS, Inc. He is the author of many books and scientific publications, and is a frequent speaker at international events. Mark Burgess may be found at www.markburgess.org, and on Twitter under the name @markburgess_osl.